Comprehensive Seismic Zonation Schemes for Regions at Different Scales

T. G. Sitharam • Naveen James
Sreevalsa Kolathayar

Comprehensive Seismic Zonation Schemes for Regions at Different Scales

Springer

T. G. Sitharam
Department of Civil Engineering
Indian Institute of Science Bangalore
Bangalore, Karnataka, India

Naveen James
Department of Civil Engineering
Indian Institute of Technology Ropar
Rupnagar, Punjab, India

Sreevalsa Kolathayar
Department of Civil Engineering
Amrita Vishwa Vidyapeetham
Coimbatore, Tamil Nadu, India

ISBN 978-3-319-89658-8 ISBN 978-3-319-89659-5 (eBook)
https://doi.org/10.1007/978-3-319-89659-5

Library of Congress Control Number: 2018942498

Printed on acid-free paper

This Springer imprint is published by the registered company Springer International Publishing AG part of Springer Nature.
The registered company address is: Gewerbestrasse 11, 6330 Cham, Switzerland

Preface

The practice of earthquake engineering involves identification, evaluation, and mitigation of seismic hazards. This book deals with the subdivision of regions into zones that have relatively similar exposure to various earthquake-related effects. The basis of seismic zonation is to model the rupture mechanism at the source of an earthquake, evaluate the propagation of waves through the earth to the top of the bedrock, determine the effects of local soil profile and thus develop a hazard map indicating the vulnerability of the area to the potential seismic hazards. Many earthquakes in the past have left many lessons to be learned which are essential to plan infrastructure and even mitigate such calamities in future. India has been facing threats from earthquakes since ancient times. Urbanization of cities is growing and expanding due to increasing migration of people to cities. The role of geological and geotechnical data is becoming very important in the urban planning of city infrastructure, which can recognize, control, and prevent geological hazards.

Chapter 1 presents an overview of earthquake hazards and the need for seismic zonation. Different levels of seismic zonation and issues associated with it are also discussed in this chapter. Chapter 2 presents the basic concepts of earthquake and seismicity highlighting theory of plate tectonics and global seismicity. The seismicity analysis of earthquake catalog, seismic source models, and preparation of seismotectonic atlas are discussed in this chapter. Chapter 3 presents the state-of the-art seismic hazard assessment using deterministic and probabilistic approaches, along with a few case studies of hazard assessment. Chapter 4 presents several site characterization techniques and case studies of site characterization at micro- and macro-levels. Chapter 5 discusses the effect of local site conditions on ground motion and methods for estimating local site effects. Chapter 6 presents the liquefaction mechanism and factors affecting the liquefaction susceptibility of the soil. The evaluation of liquefaction potential is elaborated with case studies. The chapter also discusses several schemes of mitigating the liquefaction. Chapter 7 presents integration of various hazards and several case studies of liquefaction studies in India at micro-, meso-, and macro-levels. The chapter also summarizes principles and practices of seismic zonation, along with brief guidelines for the same.

Study of seismic hazard and preparation of microzonation maps will provide an effective solution for urban planning. Seismic hazard and microzonation of cities enable engineers to characterize potential seismic vulnerability/risk that needs to be taken into account while designing new structures or retrofitting existing ones. This will even help in designing buried lifelines such as tunnels, water and sewage lines, gas and oil lines, and power and communication lines. This book can serve as a guideline for seismic zonation for practicing engineers and policy makers. The technicalities presented in the book will be highly beneficial to researchers and students.

Bangalore, India T. G. Sitharam
Rupnagar, India Naveen James
Coimbatore, India Sreevalsa Kolathayar
February, 2018

Contents

Acronyms

AHP	Analytical hierarchy process
CPT	Cone penetration test
CRR	Cyclic resistance ratio
CSR	Cyclic stress ratio
DEM	Digital elevation model
DSHA	Deterministic seismic hazard analysis
EHP	Earthquake hazard parameter
ENA	Eastern northern America
FS	Factor of safety
GIS	Geographic Information Systems
LPI	Liquefaction potential index
MARE	Mean annual rate of exceedance
MASW	Multichannel analysis of surface waves
MFS	Magnitude scaling factor
NEHRP	National earthquake hazard reduction program
NPP	Nuclear power plant
PHA	Peak horizontal acceleration
PI	Peninsular India
PSHA	Probabilistic seismic hazard analysis
SPT	Standard penetration test
V_s	Shear wave velocity
σ	Total stress
σ'	Effective stress
u_p	Pore pressure
γ_t	Threshold strain
γ_c	Cyclic strain
σ'	Effective stress
λ_M	Mean annual rate of exceedance of earthquakes with magnitude "*M*"
M	Magnitude
M_w	Moment magnitude

M_{cut}	Cutoff magnitude
a and b	Seismicity parameters
$\hat{n}_i$	Smoothed number of earthquakes in ith cell
$m_{\max}$	Maximum magnitude
$\nu(z)$	Mean annual rate of exceedance of ground motion parameter Z
V_{s30}	Average shear wave velocity for top 30 m overburden
β	Damping ratio
G	Shear modulus
$G_{\max}$	Shear modulus at very low strain
γ	Shear strain
γ_{eff}	Effective shear strain
$\gamma_{\max}$	Peak shear strain
Y_{br}	Bedrock level acceleration
FS_L	Factor of safety against liquefaction
$a_{\max}$	Surface level peak ground acceleration
σ_{v0}	Total overburden pressure
σ'_{v0}	Effective overburden pressure
r_d	Depth reduction factor
$(N_1)_{60cs}$	Corrected SPT value for energy efficiency, overburden pressure, and fine content
N_{req}	SPT value required to resist liquefaction
q_c	CPT value required to resist liquefaction
$m_{\min}$	Minimum magnitude
X_m	Rupture length
$N_i(m_o)$	Weightage factor for an individual fault
u_s	Displacement at surface
u_r	Displacement at rock
Y_s	PHA at the ground surface
F_s	Amplification factor
$W(z)$	Depth weightage factor
N_a and N_m	Number of acceleration and magnitude increment
a_i and m_i	Incremental values in acceleration and magnitude
$\Delta\lambda_{a_i,m_j}$	Incremental annual frequency of exceedance for a_i and m_j
$\text{CSR}_{eq,i}$	Cyclic stress ratio for an acceleration a_i
P_a	Atmospheric pressure
θ	Regression coefficients
R_i	Rank assigned to each category of a single theme
$R_{\min}$ and $R_{\max}$	Minimum and the maximum rating value of that theme
ϕ_{diff}	Phase difference
f	Frequency
C_f	Phase velocity
C_N	Overburden pressure correction
C_E	Hammer energy correction
C_B	Borehole diameter correction
C_S	Liner correction

C_R	Rod length correction
C_{FC}	Correction for fine contents
$S_T(\omega, \beta)$	Target response spectrum
$G(\omega)$	Power spectrum
$S_a(\omega, \beta)$	Observed response spectrum
$X(t)$	Stationary time history
$E(t)$	Non-stationary intensity envelope

Chapter 1
Introduction and Overview

1.1 Introduction

Natural calamities like earthquakes can neither be predicted nor be prevented. Although earthquakes have been reported from all parts of the world, they are mainly associated with the boundaries of the plates, which form the Earth's crust. The earth's crust is being slowly displaced at the boundaries of the plates. Differential displacements give rise to elastic strains, which eventually exceed the strength of the rocks involved, and faulting occurs. The strained rocks rebound along a fault under the elastic strains until the strain is partly or wholly dissipated.

Past earthquakes have demonstrated that Indian subcontinent is highly vulnerable to earthquake hazards. It has been estimated that about 59% of the land area of the Indian subcontinent has potential risk from moderate to severe earthquakes [237]. Major earthquakes in the last 20 years such as Khillari (30th September 1993), Jabalpur (22nd May 1997), Chamoli (29th March 1999) and Bhuj (26th January 2001) earthquakes have resulted in more than 23,000 deaths and extensive damage to infrastructure [237]. Although it is well known that the major earthquake hazard prone areas in India are the Himalayan region (inter-plate zone) and the northeast region (subduction zone), the seismicity of Peninsular India cannot be underestimated. Many studies [102, 275, 304] have proved that the seismicity of Peninsular India is significantly high and may lead to earthquakes of sizeable magnitude. This necessitates a seismic zonation for the country, as well as various regions in it. Seismic zonation is the first step towards an effective earthquake risk mitigation study.

Seismic zonation is a process in which a large region is demarcated into small zones based on the levels of earthquake hazard. Seismic Zonation is delineation of regions having diverse potentials for hazardous effects from impending earthquakes. Within a seismic zone, the seismicity rate is assumed to be consistent. Seismic

T. G. Sitharam et al., *Comprehensive Seismic Zonation Schemes for Regions at Different Scales*, https://doi.org/10.1007/978-3-319-89659-5_1

zonation is a process in which a large region is demarcated into small zones based on the levels of earthquake hazard. A seismic zoning map for engineering use is a map that defines the levels of ground motions for earthquake-resistant design, and thus it differs from a seismicity map, which provides only past earthquake data and seismic sources. The task of seismic zoning is multidisciplinary that involves the contribution from geologist, seismologist, geotechnical, earthquake, and structural engineers. Seismic zonation can act as a roadmap for designers and policy makers to identify the regions with potential seismic threat.

1.2 Earthquake Damages

Earthquakes do not pose great danger to mankind directly as public cannot be shaken to death by an earthquake. The casualties due to earthquake happen because of associated hazards and infrastructure failures.

1.2.1 Ground Shaking

The first main earthquake hazard is the effect of ground shaking. Buildings can be damaged by the ground shaking or by the ground subsidence where ground settles to a different level during the earthquake. The extent of ground shaking will depend upon the scale of an earthquake, distance from the epicenter, and nature of the material.

1.2.2 Liquefaction

Liquefaction is the mixing of soil and groundwater during the shaking because of an earthquake because of which soil loses its shear strength and behaves like a fluid. If liquefaction occurs under a building, it may sink into the ground. Liquefaction is a phenomenon seen in sandy soils with groundwater near the surface.

1.2.3 Landslides

The ground shaking may cause instability of slopes that can lead to landslides, mudslides, and avalanches, which can damage buildings located nearby.

1.2.4 Ground Displacement

If a structure (a building, road, etc.) is built across a fault, the ground displacement during an earthquake could seriously damage or rip apart that structure.

1.2.5 Flooding

An earthquake can break dams which will lead to flooding of the area downstream, damaging buildings, and sweeping away people.

1.2.6 Tsunami

A tsunami is a huge wave caused by an earthquake under the ocean. Tsunamis can be tens of feet high when they hit the shore and can do massive damage to the coastline. Seiches occur on lakes that are shaken by the earthquake and are usually only a few feet high, but they can still flood the surroundings.

1.2.7 Fire

These fires can be started by broken gas lines and power lines during an earthquake. They can be a serious problem, especially if the water lines that feed the fire hydrants are broken, too. After the Great San Francisco Earthquake in 1906, the city burned for 3 days leading to massive destruction of the city leaving 250,000 people homeless.

Most of the hazards to people come from man-made structures themselves and the shaking they receive from the earthquake. The real dangers to people are being crushed in a collapsing building, drowning in a flood caused by a broken dam or levee, getting buried under a landslide, or being burned in a fire.

1.3 Seismic Zonation

Seismic zonation is a process of dividing a large region into small zones based on the expected level of earthquake hazard. Seismic zonation helps to identify vulnerable regions and also provide necessary outputs for the earthquake resistant design. Hence it is very much required in the modern world in order to minimize the casualty and economic losses during an earthquake. The large and rapidly growing urban seismic risk, particularly in developing countries like India is a problem that needs to be quickly solved. This can be done through a comprehensive seismic microzonation of the area. The ISSMGE Technical Committee for Geotechnical Earthquake Engineering proposed three grades of zoning along with the accepted scales of mapping [144]. The scale of zonation depends mainly on the available database and on the quality of the zonation map required. Several inputs regarding seismicity, geology, geomorphology, and geotechnical characteristics are needed for doing seismic microzonation of an area.

1.4 Need for Seismic Zonation

Seismic microzonation is the first step in earthquake risk mitigation study and requires multidisciplinary approach with major contributions from the fields of geology, seismology, geophysics, geotechnical, and structural engineering. This is very important to identify the tectonic and geological formations in the study area which is essential for determining the seismic sources and also for establishing a realistic earthquake hazard models for the investigation.

The very important issue affecting the applicability and the feasibility of any microzonation study is the reliability of the parameters for doing it. It is very clear based on the earthquake damage and strong motion records that there are numerous source and site factors i.e., near field effects, directivity, duration, focusing, topographical and basin effects, etc. which are important in assessing ground motion characteristics. The national seismic zoning maps which are prepared in small scales like 1:1,000,000 or less neglecting all the above factors and they do not consider the geological and geotechnical site conditions.

Seismic microzonation involves a very detailed field investigation to evaluate the hazard. It is very effective in delineating the spatial variations in the seismic hazard. They are also useful to evaluate the risk scenarios in the study area. This has been the most widely used method to map earthquake hazard at local scales which may incorporate a wide variety of information, including seismic response of different surface geological formations, liquefaction potential, topographic amplification of seismic waves, landslides, tsunamis, etc. Seismic microzonation maps are very useful in urban planning because they help to predict the impact of future earthquakes and can also be used to locate key facilities like hospitals, fire stations, emergency operation centers, etc. Microzonation studies are also very useful to save the heritage and important structures from future major earthquakes.

The Turkish codes insist that the seismic microzonation has to be done in vulnerable cities/regions with a population of 30,000 or more. However, since the population in India is very high, the microzonation need to be carried out for the cities based on population and macro hazard maps available in the country. For example, microzonation can be taken up for the urban centers with a population of 50,000 to one lakh in an area and which falls in seismic zone five. While selecting the regions for microzonation, the population of the region, its socio-economic characteristics, and the strategic importance, etc. also need to be considered.

1.5 Level of Seismic Zonations and Methodology

Seismic zonation is generally carried out at three different levels based on the aerial extent of the region, the importance of the site and the population. They are micro-level, meso-level, and macro-level. The macro-level zonation is generally carried out for large landmass such as a state or a country. The earthquake hazard parameters used for macro-level zoning are generally evaluated with less reliability. The typical example of a macro-level zonation is the seismic zonation map of India prepared by IS-1893, where the entire India is demarcated into four seismic zones based on past seismicity and tectonic conditions. Generally the macro-level seismic zonation is carried out based on peak horizontal acceleration (PHA) estimated at bedrock level without giving emphasis on the local soil conditions. Seismic zonation at the meso-level is carried out for cities and urban centers with a population greater than 500,000. The earthquake hazard parameters, for the meso-level zonation are evaluated with greater degree of reliability, compared to the macro-level zoning. The micro-level zonation is carried out for sites which host critical installations such as nuclear power plants (NPPs). As the NPPs are considered as very sensitive structures, the earthquake parameters, for the micro-level zonation of the NPP sites are estimated with a highest degree of reliability. The local soil conditions and site effects are properly counted for carrying out the micro as well as the meso-level zonation. Several researchers have carried out meso-level zonation considering the effects of all major earthquake hazards such as PHA, site amplification, and liquefaction [223, 235, 328].

The local site effects during earthquakes are related to geotechnical characteristics such as amplification, liquefaction, land slide, mudflow and fault movements. To assess these geotechnical characteristics three grades of approaches to zonation was suggested by ISSMGE Technical Committee for Geotechnical Earthquake Engineering along with the accepted scales of mapping (Table 1.1). The scale of zonation depends mainly on the available database and on the quality of the zonation map required. Several inputs regarding seismicity, geology, and geotechnical characteristics are needed for doing seismic microzonation of an area. The national seismic zoning maps are mostly at small scales such as 1:1,000,000 or less and are mostly based on seismic source zones defined at similar scales. However, seismic microzonation for a city/town requires 1:5000–1:25,000 scale studies and needs to be based on seismic hazard studies at similar scales. The country wide macrozonation maps which are produced by national experts will go through a careful review process. However it may not be possible to adopt such a review system for the large number of seismic microzonation studies in the country. Geological Survey of India maps are available at 1:50,000 scale with regard to base map, contours, geomorphology, geology, etc. It is advisable to explore these maps for microzonation studies immediately. For some important mega cities this exercise can be repeated at large scales like 1:10,000. This is presently being done for National Capital Region Delhi. One possible solution for this scale incompatibility is to increase the scales of seismic macrozonation maps

Table 1.1 Three levels of zonation (modified from [144])

Geotechnical phenomenon	Grade-1	Grade-2	Grade-3
Ground motions	Historical earthquakes and existing information geological maps interviews with local residents	Microtremor simplified geotechnical	Geotechnical investigation ground response analysis
Slope stability	Historical earthquakes and existing geological and information geomorphological maps	Air photos and remote sensing field studies vegetation and precipitation data	Geotechnical investigation analysis
Liquefaction	Historical earthquakes and existing information geological and geomorphological maps	Air photos and remote sensing field studies interviews with local residents	Geotechnical investigation analysis
Accepted scale of mapping	1:1,000,000 to 1:50,000	1:100,000 to 1:10,000	1:125,000 to 1:5000
Grid size for testing geophysical survey and bore hole	Homogeneous sub-surface—2 km × 2 km–5 km × 5 km Heterogeneous Sub-surface—0.5 km × 0.5 km–2 km × 2 km	Homogeneous sub-surface—1 km × 1 km–3 km × 3 km Heterogeneous Sub-surface—0.5 km × 0.5–1 km × 1 km	Homogeneous sub-surface—0.5 km × 0.5 km–2 km × 2 km Heterogeneous Sub-surface—0.1 km × 0.1 km 0.5 km × 0.5 km

steadily with the accumulation of geological and seismological data as implemented in USA and Japan. Geological formations, local site classification, equivalent shear wave velocity, spectral acceleration, spectral amplification, and their variation are some of the parameters studied during a seismic microzonation studies. A consistent approach has to be implemented to assess each parameter with respect to all other parameters. The objective of seismic zonation is to establish a seismic hazard map at a scale of 1:10,000 taking into account earthquake source and local site conditions. Thus estimation of the earthquake induced forces and their variation in the investigated area are the main target in seismic microzonation.

1.6 Issues Related to Seismic Zonation

The government of India has already initiated the seismic zonation of major cities in India. However, the micro-level zoning cannot be performed for all these cities as very large resources are required for an extensive geotechnical and geophysical investigation and site effect estimation. Hence there is a need to define the level of zonation to be adopted for various study areas in the country. As there are many methodologies available for the site characterization and site effect estimation, the suitability of these methodologies for each levels of zoning also needs to be assessed in order to optimize the resources for carrying out seismic zonation.

Even though the above definitions and descriptions are available for various levels of zonation, the key issue lies in the adoption of the suitable one for a given region. There are only a few guidelines available regarding the use of a particular level of zonation for a given study area. Based on the recommendation of the disaster management authority, the government of India has initiated the seismic zonation of all major cities in India. As it is evident that large resources are required in order to carry out seismic site characterization and site effect estimation, both the micro and meso-level zonations cannot be carried out for all these cities. Hence there is a need to propose appropriate guidelines to define the suitability of each level zonation for various regions in the country. Moreover, there are many methodologies available for site characterization and estimation of site effects such as site amplification and liquefaction. The appropriateness of these methodologies for various levels of seismic zonations also needs to be assessed in order to optimize use of resources for seismic zonation. Hence, in this book, appropriate techniques for site characterization and earthquake hazard estimation for regions at different scale levels were determined. Using the appropriate techniques, the seismic zonation was carried out both at the micro and macro-level, incorporating all major earthquake hazards.

The zonation shall be graded based on the scale of the investigation and details of the studies carried out. The technical committee on earthquake geotechnical engineering (TC4) of the International society of soil mechanics and foundation engineering [144] states that the first grade (Level I) map can be prepared with scale of 1:1,000,000–1:50,000 and the ground motion to be assessed based on the historical earthquakes and existing information of geological and geomorphological maps. If the scale of the mapping is 1:100,000–1:10,000 and ground motion is assessed based on the microtremor and simplified geotechnical studies, then it is called second grade (Level II) map. In the third grade (Level III) map ground motion is to be assessed based on the complete geotechnical investigations and ground response analysis with a scale of 1:25,000–1:5000.

The existing macro-level zonation map of India, used for the design practice is developed by BIS-1893 [28]. However the major drawback of the seismic zonation map of India developed by BIS-1893 [28] is that it is based on the past seismic activity and not on scientific seismic hazard analysis. The BIS code always subjected to revision only after major earthquake. Such a zonation map is not appreciated in an engineering point of view and hence it is desirable to carry out macro-level seismic zonation based on parameters such as peak horizontal acceleration (PHA), which can be estimated scientifically. A macro-level zoning refers to the zonation of a large region such as a state or a country. Several studies have been carried out for macro-level seismic hazard assessment and zoning based on the peak horizontal acceleration (PHA) at the bedrock [25, 149, 260, 376]. However still the influence of site conditions is not considered in the macro-level zoning. Similarly, the micro-level zonations are also required especially for the NPP sites. As there are lot of constraints for the selection of a site to host NPP, seismic zonation helps to identify the least hazardous sector within the site where the critical structures of NPP can be sited. Seismic zonation of existing NPP site is also required for assessing the seismic vulnerability and risk associated with various structures of NPP. Hence in the present study an attempt has been made to review the feasibility of various methodologies for site characterization and site effect estimation and thus carry out the seismic zonation, both at the micro and macro-level using the appropriate methodology. The micro-level zonation was carried out for the Kalpakkam NPP site, while the macro-level zonation was performed for the state of Karnataka. As the Kalpakkam NPP site area does not lie within the state of Karnataka, the results of the macro-level study were validated with the existing meso-level study carried out for the Bangalore city. Based on this, recommendations have been made regarding the suitability of various methodologies for different level of zonation.

1.7 Global Trends

The first attempt of seismic microzonation of any urban area, i.e. an industrial as well as population center was carried out in the city of Yokohama, Japan in 1954 considering various zones, corresponding soil conditions and design seismic coefficients for different types of structures located in that different zones. Subsequently, in view of the immense usefulness of microzonation studies were conducted in few earthquake prone areas of the world [7, 14, 64, 89, 91, 199, 208].

Slob et al. [338] presented a technique for microzonation for the city of Armenia in Colombia. In this study they used a 3D layer model in GIS, combined with a 1D calculation of seismic response using SHAKE to get the spatial variation in seismic response which was checked with the damage assessment of Armenia. They concluded that the amplification was high in the range of frequency of 5 Hz for houses with 2 stories, it became true after the earthquake, in which low rise building experienced more damage than high rise building. Topal et al. [357] considered various parameters for microzonation such as geological, geotechnical, seismotectonic, and hydrogeological conditions and on the basis of these, four different zones were proposed for the Yeneshir, an urban area in Turkey. Ansal et al. [14] adopted a probabilistic approach in a microzonation study for the city of Siliviri, Turkey and estimated the local site effects on the basis of the available borehole data and laboratory results on collected samples from the area. For site characterization the average shear wave velocity was used, that was determined from seismic refraction tests.

Seismic microzonation in India is recognized as a tool for disaster mitigation and for providing guidelines to the construction engineers. Some states of country have made it mandatory for the new constructions to be declared seismically safe and for this prior approval of the experts is necessary. After the devastating earthquake in Gujarat in 2001, the Government of India has paid serious attention to seismic microzonation and it has accepted its importance as a guiding tool in land use planning and safe construction practices to avoid the loss from the future earthquakes. As a result of this, seismic microzonation of various urban areas, like Delhi, Jabalpur, Chennai, Bangalore, Lucknow, Ahmedabad, etc., are being carried out by different researchers in the country.

Chapter 2
Earthquake and Seismicity

2.1 Plate Tectonics

The science of earthquakes requires knowledge of plate tectonics, a theory that describes the movement of major and many smaller geological deformations that make up the earth's crust. These interlocking crustal plates cover the surface of the earth, and are moved by radioactivity-driven currents in the earth's mantle. The heavier ocean-floor plates, expanding from central ridges, meet, and sink under, the lighter continental plates. As the plates meet, they often lock up, instead of sliding by each other. This builds up huge stresses; when the plates suddenly unlock with one powerful jerk the release of this tension is what causes the earthquake. A subduction earthquake is caused when one edge of a plate is forced down under another edge; it is considered to be potentially the most dangerous earthquake.

The surface of the earth is made up of several wide, thin, and rigid plate-like blocks (Fig. 2.1). These are constantly rubbing against each other; and sometimes, their rubbing is strong enough to shake up the earth, release tremendous amounts of energy, and cause havoc. The current movement of the Indian plate is estimated to be around 50 mm/year. Of course, in our day to day lives, this kind of movement is negligible. However, a plate beneath the earth's surface, moving at this rate for a long enough time, can have a strong influence on the structure of the earth itself. The earth's crust is made up of seven major plates and different minor plates and these plates are moving relative to each other. Based on the types of relative movement of the plates, the plate boundaries are divided into three types: spreading ridge boundaries, subduction zone boundaries, and transform fault boundaries. The earthquake activity is very high along the subduction and transform plate boundaries. Based on the location of earthquakes, it can be divided into

T. G. Sitharam et al., *Comprehensive Seismic Zonation Schemes for Regions at Different Scales*, https://doi.org/10.1007/978-3-319-89659-5_2

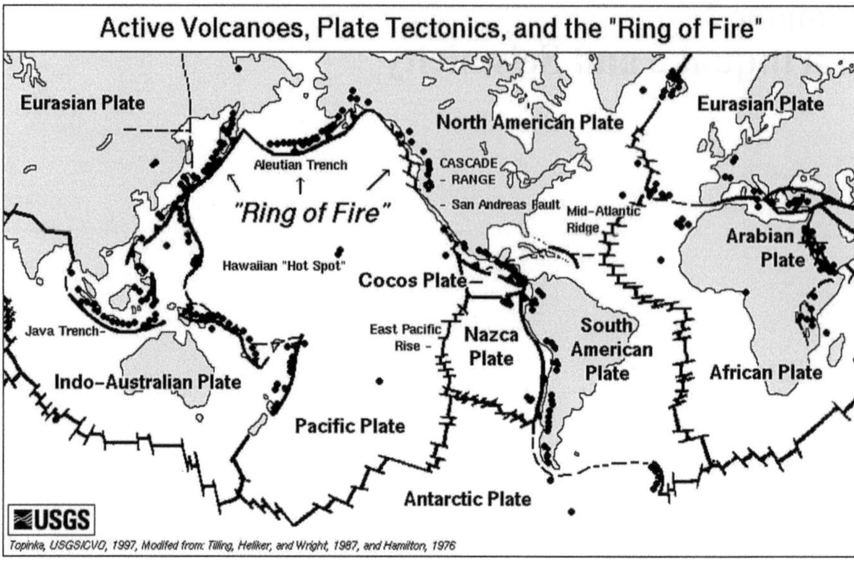

Fig. 2.1 Major tectonic plates (source: USGS)

two types—interplate earthquakes and intra plate earthquakes [282]. The main difference between the intra plate earthquakes and the inter plate earthquakes are:

- The recurrence time will be more for intra plate earthquakes
- The faults are rarely visible at the surface for intra plate earthquakes
- The energy released will be more than that of an inter plate earthquake
- The seismic energy dissipates very slowly and the seismic waves travel a longer distance for intra plate earthquakes

2.1.1 Faulting of Rocks

Faults are localized areas of weakness in the surface of the Earth, sometimes the plate boundary itself. A fault can be a crack in the earth's crust along which rock on one side is displaced relative to that on the other side (Fig. 2.2). Movements in the plates create stresses in the plate and if the stress reaches ultimate strength of the rock, the rock breaks and this results in sudden release of stored strain energy. This breakage of strata over large areas is known as faulting. Different variety of faults will result i.e., strike slip fault, normal fault and reverse or thrust faults depending upon stress causing the faulting. As the broken surface is planar in nature, the fault plane designated with three parameters, namely strike, dip, and dip direction. Movement along the fault of the two broken surfaces (usually called hanging wall and footwall) is called slip and there are several varieties of slips possible.

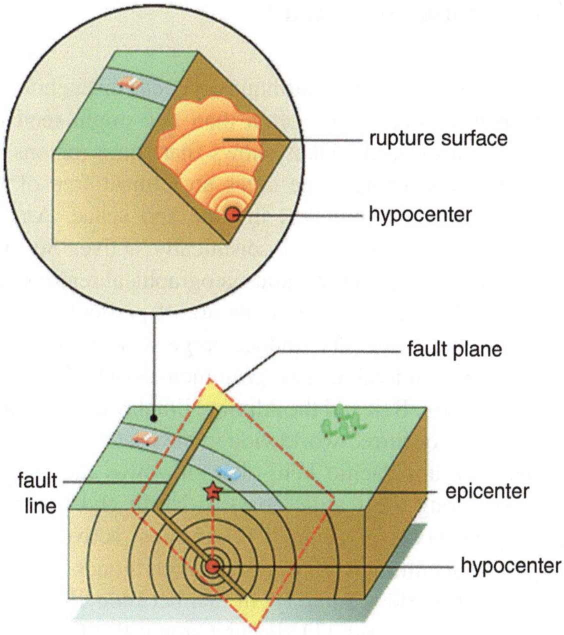

Fig. 2.2 Rupture of rock along a fault (source: USGS)

2.1.2 Earthquakes: Elastic Rebound Theory

Earthquake is associated with sudden rapid shaking of the earth causing fracture in the earth's crust, leading to release of strain energy stored in rocks which spread in all directions causing ground shaking. As stated, the plates of the earth are in constant motion and plate tectonics indicates that the majority of their relative movements occur near the boundaries. As the relative movement of the plates occur, elastic strain energy is stored in the material near the boundary. When the shear stress reaches the shear strength of the rock along the fault, the rock faults and the accumulated strain energy is released. Rupture of the rock will release the stored energy explosively, partly in the form of heat and partly in the form of the stress waves that are felt as earthquakes. The theory of elastic rebound describes the process of successive build up and release of strain energy in the rock. The elastic rebound theory explains the spreading of energy during earthquakes. When plates on opposite sides of a fault are subjected to force and shift, they accumulate energy and slowly deform. As the developed stresses exceed the internal strength of the rock, a sudden movement occurs along the fault, releasing the accumulated energy, and the rocks snap back to their original undeformed shape.

This theory was discovered by making measurements at a number of points across a fault. Prior to an earthquake it was noted that the rocks adjacent to the fault were bending. These bends disappeared after an earthquake suggesting that the energy stored in bending the rocks was suddenly released during the earthquake.

2.2 Global Seismicity

A very large number of earthquakes occur throughout the world every year; in fact, earthquakes occur more often than one might tend to believe. However, spatial distribution of earthquakes shows that some regions have more earthquakes than other regions, while large areas are almost free of seismicity. Seismicity is the distribution of earthquakes in time and space. Any region, which has frequent earthquakes, is considered seismically active. Seismicity is concentrated along certain narrow, semicontinuous geographical regions called seismic belts. These are shown in Fig. 2.3. Seismic belts are of particular interest as frequent earthquakes occur in these regions, induce large-scale damage repeatedly, and make large populations vulnerable. Two prominent seismic belts can be identified on the globe; Circum Pacific Belt and the Alpine-Himalayan Belt. Table 2.1 lists major destructive earthquakes occurred worldwide in the past.

The Circum Pacific belt, also known as the ring of fire, is long and narrow. It exists along the Pacific coast of North and South America and continues into the Pacific coast of Asia. It is the most active of all seismic belts and has the largest concentration of devastating earthquakes. It contributed more than three quarters of world seismicity; in fact between 1904 and 1952, it gave off 75.6% of global seismic energy [113]. The Circum Pacific belt is very complex and includes special topographic features such as island arcs, oceanic trenches, and mountain ranges. It has intermediate and deep focus earthquakes, together with shallow focus earthquakes.

The Alpine-Himalayan belt is the next most active belt. It contributed 22.1% of seismic energy given off on the globe between 1904 and 1954. This seismic belt is

Fig. 2.3 Map showing worldwide seismic hazard (source: USGS)

Table 2.1 Lists of 20 largest earthquakes in the world (source: [80, 154, 164–167, 198, 246, 247])

Sl. no.	Mag	Location	Alternative name	Date (UTC)	Time UTC	Lat	Long
1	9.5	Bio-Bio, Chile	Valdivia earthquake	5/22/196	19:11	38.14°S	73.41°W
2	9.2	Southern Alaska	1964 Great Alaska Earthquake, Prince William Sound Earthquake, Good Friday Earthquake	3/28/1964	3:36	60.91°N	147.34°W
3	9.1	Off the West Coast of Northern Sumatra	Sumatra-Andaman Islands Earthquake, 2004 Sumatra Earthquake and Tsunami, Indian Ocean Earthquake	12/26/2004	0:58	3.30°N	95.98°E
4	9.1	Near the East Coast of Honshu, Japan	Tohoku Earthquake	3/11/2011	5:46	38.30°N	142.37°E
5	9.0	Off the East Coast of the Kamchatka Peninsula, Russia	Kamchatka earthquake	11/4/1952	16:58	52.62°N	159.78°E
6	8.8	Offshore Bio-Bio, Chile	Maule Earthquake	2/27/2010	6:34	36.12°S	72.90°W
7	8.8	Near the Coast of Ecuador	1906 Ecuador—Colombia Earthquake	1/31/1906	15:36	0.96°N	79.37°W
8	8.7	Rat Islands, Aleutian Islands, Alaska	Rat Islands Earthquake	8/15/1950	14:09	28.36°N	96.45°E
9	8.6	Eastern Xizang-India Border Region	Assam, Tibet	8/15/1950	14:09	28.36°N	96.45°E
10	8.6	Off the West Coast of Northern Sumatra		4/11/2012	8:39	2.33°N	93.06°E
11	8.6	Northern Sumatra, Indonesia	Nias Earthquake	3/28/2005	16:10	2.09°N	97.11°E
12	8.6	Andrean of Islands, Aleutian Islands, Aleutian Islands,		3/9/1957	14:23	51.50°N	175.63°W
13	8.6	South of Alaska	Unimak Island Earthquake, Alaska	4/1/1946	12:29	53.49°N	162.83°W
14	8.5	Banda Sea		2/1/1938	19:04	5.05°S	131.61°E
15	8.5	Atacama, Chile	Chile-Argentina Border	11/11/1922	4:33	28.29°S	69.85°W
16	8.5	Kuril Islands		10/13/1963	5:18	44.87°N	149.48°E

(continued)

Table 2.1 (continued)

17	8.4	Near the East Coast of Kamchatka Peninsula, Russia	Kamchatka, Russia	2/3/1923	16:02	54.49°N	160.47°E
18	8.4	Southern Sumatra, Indonesia		9/12/2007	11:10	4.44°S	101.37°E
19	8.4	Near the Coast of Southern Peru	Arequipa, Peru Earthquake	6/23/2001	20:33	16.27°S	73.64°W
20	8.4	Off the East Coast of Honshu, Japan	Sanriku, Japan	3/2/1933	17:31	39.21°N	144.59°E

more diffuse than the Circum Pacific belt. Topographic features associated with this belt are mountain ranges on continents and island arcs and deep trenches in oceans. It includes the mountainous regions of Alps in Europe, Zagros in Iran, Sulaiman and Kirthar ranges in Pakistan, Hindu Kush and Pamir regions, the Himalayas in Asia, and extends toward the East Indies, via the Arakan Yoma mountain ranges and continues eastward into Indonesia and Philippines. It includes the mountain ranges that radiate from the Pamir knot, such as Karakoram, Kunlun, Altyn Tagh, and those that stretch into Tibet, China, and Mongolia.

Besides the Circum Pacific Belt and the Alpine-Himalayan Belt, other regions of reduced seismicity also exist on the globe. These comprise of mid-oceanic ridges, continental rifts, marginal areas, regions of old seismicity, and stable masses. Regions of old seismicity refer to pre-Cambrian shields of Africa, India, Siberia, Fenoscandia, Australia, Canada, and Brazil.

2.3 Indian Seismicity

The earthquake zoning map of India divides India into four seismic zones (Zone II to V) as shown in Fig. 2.4, unlike its previous version which consisted of five or six zones in the country. According to the present zoning map, Zone-V expects the highest level of seismicity whereas Zone-II is associated with the lowest level of seismicity.

The tectonic framework of the Indian subcontinent is diverse with respect to space and time and hence complex. The rapid drifting of the Indian plate towards the Himalayas in the northeastern direction with a high velocity along with its low plate thickness [185] might be the cause for the high seismicity of the Indian region. When continents converge, large amounts of shortening and thickening take place, like in the Himalayas and Tibet. Due to this massive collision, the Himalayas were formed and large numbers of earthquakes happen in the region. A similar process, involving the Indian Plate and the Burmese micro-plate, results in earthquakes in

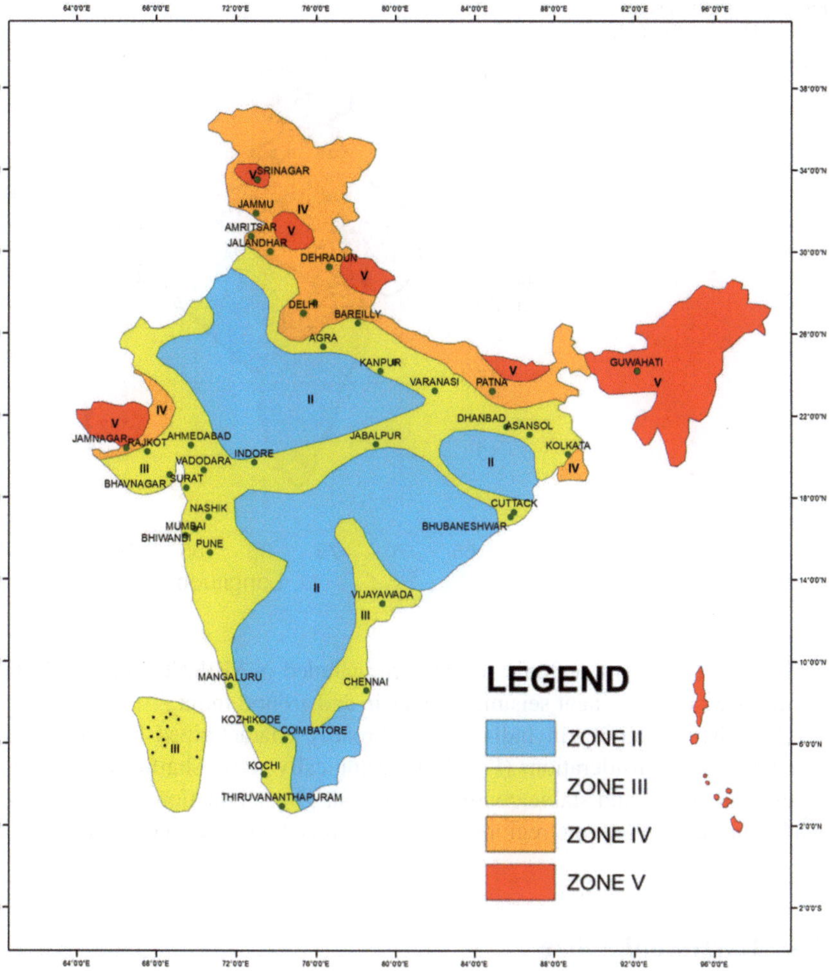

Fig. 2.4 Latest seismic zonation map of India [28]

the Andaman & Nicobar Islands. The plate boundary areas along the Himalayas and North East India are characterized by a very high level of seismicity. There are earthquakes occurring within the Indian shield region, in the Indian peninsula, and in adjoining parts of the Arabian Sea or the Bay of Bengal, which are intraplate earthquakes.

Geological Survey of India published Seismotectonic Atlas, SEISAT (2000) with details of linear seismic sources in India and adjoining areas with all available data related to earthquakes. It is a compilation of multi-thematic database comprising of 43 maps covering India and adjacent regions of neighboring countries on 1:1 million scale. Various details regarding geophysical, structural, seismicity, and geothermal

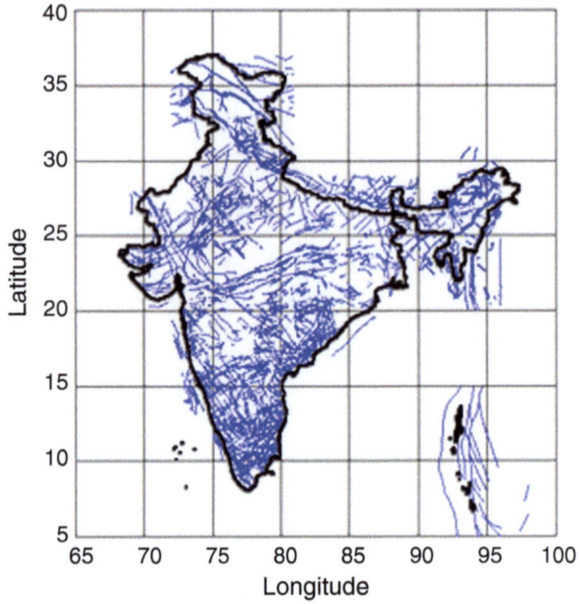

Fig. 2.5 Linear seismic sources identified in India [316]

data relevant to seismotectonic activity are included in SEISAT. Figure 2.5 shows the digitized map of linear seismic sources in and around India.

The seismic activity in India can be broadly characterized by three general seismotectonic considerations (Fig. 2.6): Tectonically active shallow crustal region, subduction zones, and stable continental region. The subduction zone earthquakes can be further divided into regions with intraslab and interface earthquakes.

2.4 Earthquake Size

The news reports of earthquakes refer to an earthquake with a certain magnitude, or that with a certain number on the Richter scale. For example, 2001 Bhuj earthquake was a magnitude 7.7 earthquake and 2015 Nepal earthquake was a magnitude 7.7 earthquake. Most of us understand that the higher the number, the larger the quake; but how are these numbers arrived at, and is this the only way to measure an earthquake? The size of an earthquake depends on the size of the fault and the amount of slip on the fault. Scientists use an instrument called the seismograph to measure the magnitude of an earthquake. Seismographs record the movement of vibrations beneath the earth, as zig-zag traces on a sheet of paper. Some seismographs are sensitive enough to detect earthquakes happening anywhere on earth.

Earthquake size can be expressed qualitatively (non-instrumental) or quantitatively (instrumental). It is commonly expressed in terms of Intensity or Magnitude. Intensity is a measure of how strong the earthquake feels to an observer. It is the

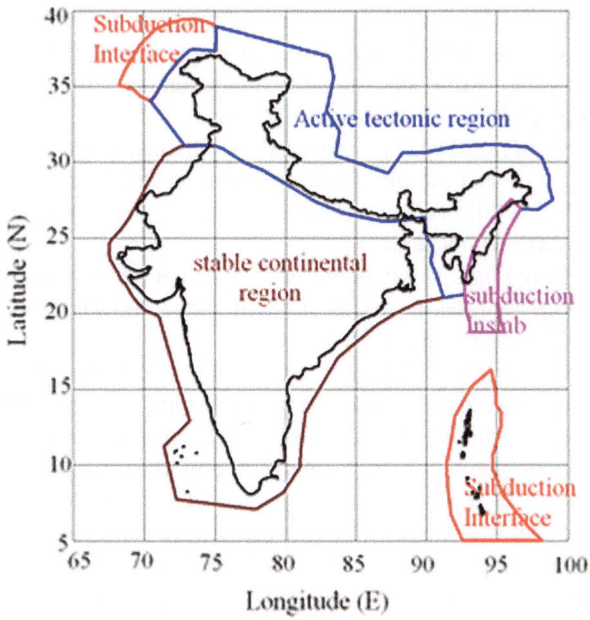

Fig. 2.6 Tectonic provinces in and around India [178]

qualitative assessment of the kind of damage done by an earthquake. Intensity depends on distance to epicenter, strength of earthquake, and local geology. It is determined from the intensity of shaking and the damage from the earthquake. Magnitude is the quantitative measurement of the amount of energy released by an earthquake. Such a quantitative measure is needed to compare the size of earthquakes worldwide, which is independent of the density of population and the type of construction. The magnitude of an earthquake is a very important parameter in seismic hazard analysis and this has been reported using different scales. Some of the important earthquake magnitude scales are described below.

1. *Intensity scale* (I): Intensity is a qualitative measurement of earthquake size as it reflects the strength of an earthquake from the observer's point of view. Intensity value also indicates the kind of damage done by an earthquake and depends neither on the seismological attributes of the earthquake nor on local geology. As seen, the magnitude of an earthquake gives an indication of the total energy released but it cannot give a clear picture of the damaging effects. The intensity scales is a measure of the damaging effect of an earthquake produced at a given location. Moreover this scale can be used to indicate the size of earthquakes for regions where there are no instrumental records are available. There are several intensity scales developed by various researchers such as Mercalli-Cancani-Selberg (MCS), Modified Mercalli (MM), Medvedev-Sponheuer-Karnik (MSK), Japanese Meteorological Agency (JMA), Rossi-ForelIntensity scale.

2. *Local magnitude* (M_L): or the Richter scale magnitude proposed by Richter [283] for measuring shallow earthquakes. A magnitude of 3 in the Ritcher scale magnitude corresponds to an earthquake, producing an amplitude of $A = 1$ mm on a seismograph kept at a distance of 100 km.
3. *Surface wave magnitude* (M_S): proposed by Gutenberg and Richter [111], for measuring size of shallow and distant earthquakes (focal depth less than 70 km and farther than 1000 km). The surface wave magnitude is based on the amplitude of Rayleigh wave with a period of 20 s.
4. *Body wave magnitude* (m_b): proposed by Gutenberg and Richter [114] for measuring deep focus earthquake. It is based on the maximum amplitude of first few cycles of P waves with a period of 1 s.
5. *Moment magnitude* (M_W): The above magnitude scales (M_L, M_S, m_b) measure the magnitude of earthquake based on the amplitude of motion. Moreover these magnitude scales become saturated at different levels for large earthquakes [395]. In order to overcome these shortcomings, the moment magnitude (MW) scale was introduced. In this scale the magnitude of earthquake is measured with respect to the energy released. During an earthquake the rupture along a fault will create equal and opposite forces and it will create a force couple. The moment magnitude of an earthquake can be calculated as Eq. (2.1).

$$M_W = \frac{2}{3}\log_{10} M_o - 10.7 \tag{2.1}$$

$$M_o = \mu A D \tag{2.2}$$

where M_o—seismic moment, μ—modulus of rigidity of the rock mass, A—area of the fault along the fault, D—average displacement, where M_o is in dyne-cm and calculated using Eq. (2.2). In the above equation the value of M_o should be used in dyne-cm. Nowadays this is the most widely used magnitude scale for most of the scientific applications.

As shown in Fig. 2.7 all other magnitude scales except M_W saturate for large earthquakes. It is apparent that m_b begins to saturate at $m_b = 5.5$ and fully saturates at 6.0, whereas, M_S does not saturate until approximately $M_S = 7.25$ and is fully saturated by 8.0. M_L begins to saturate at about 6.5. It is desirable to have a magnitude measure that does not suffer from this saturation.

2.4.1 Magnitude Conversion

It is a pre-requisite for complete earthquake catalogue to have a uniform magnitude scale for denoting the size of earthquakes, so that a reliable parameterization of the magnitude distribution which is homogeneous and complete with respect to time and size is used in hazard analysis. The earthquake data are generally

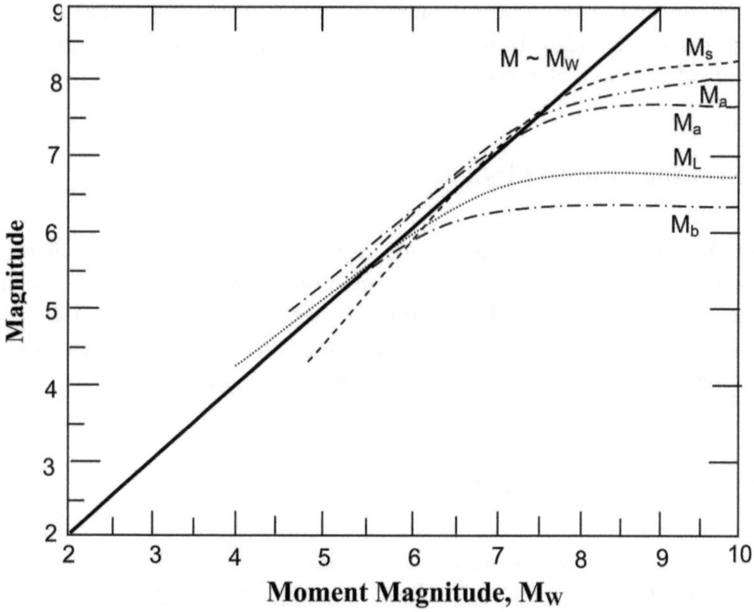

Fig. 2.7 Comparison of moment magnitude scale with other magnitude scales (after [126])

available in different magnitude scales like body wave magnitude (m_b), surface wave magnitude (M_S), local magnitude (M_L), moment magnitude (M_W), and the earthquake intensity scale (I). Unfortunately, many of the magnitude scales are all limited by saturation toward large earthquakes with $m_b > 6.0$, $M_L > 6.5$, and $M_S > 8.0$. The existence of different magnitude scales necessitates the conversion of these magnitude scales to a single magnitude scale for the analysis purposes. The moment magnitude M_W [166] can represent the true size of earthquakes because it is based on seismic moment, which in turn is proportional to the product of the rupture area and dislocation of an earthquake fault [5]. M_W is defined as

$$M_W = \frac{2}{3}\log_{10} M_o - 6.05 \tag{2.3}$$

where M_o is the scalar seismic moment in $N-m$. The homogenization of earthquake catalog involves expressing the earthquake magnitudes in one common scale. Practical problems, such as seismic hazard assessment, necessitate the use of homogenized catalogue. Since M_W does not saturate, this is the most reliable magnitude for describing the size of an earthquake [303]. As the moment magnitude (M_W) scale is the most advanced and widely used magnitude scale, it is recommended to convert the original magnitudes of earthquake events in different time periods to unified M_W magnitudes. Several relations were proposed by different researchers to convert different magnitude scales to M_W ([107, 126, 155, 173, 175, 241, 256, 262, 303, 354]; among many others). The relations between different magnitude scales will depend on the observation errors, source characters such as stress drop, fault geometry, etc. [126]. Hence it is always advisable to use the region specific magnitude conversion relations [196].

2.5 Seismic Source Models

One of the major steps in the seismic hazard analysis is the identification of vulnerable seismic sources. A seismic source model primarily defines the spatial characteristics of a source and the earthquake distribution within it. The different types of seismic sources considered in general are linear seismic sources, point sources, Gridded Seismicity source model, and areal sources.

2.5.1 Linear Source Model

A linear source model comprises mainly of faults and lineaments. A lineament is a linear feature on the earth's surface whose parts align in a straight or slightly curved and is distinctly different from adjacent features. One of the best documents listing the linear seismic sources in India and adjoining areas is the Seismotectonic Atlas [316] published by the Geological Survey of India (GSI). The seismotectonic atlas [316] was prepared after extensive studies using remote sensing technique and by geological explorations. The SEISAT maps [316] are available in A_0 size sheets with 1:1,000,000 scale and each map covers an area of $3° \times 4°$. SEISAT contains the details of the faults, lineaments, and shear zones in addition to the geological features in India and adjoining areas. This has been taken as an authentic reference manual for identifying the seismic sources by various researchers like [146] for Delhi, [235] for microzonation of Sikkim Himalayas, [273] for Mumbai, [33] for Chennai, and [10] for Bangalore.

Apart from Seismotectonic atlas, the major lineaments in a given study area also can be mapped using satellite data such as Indian Remote Sensing Satellite (IRS)-1D, Wide Field Sensor (WiFS), and Landsat Multi Spectral Scanner (MSS)/Thematic Mapper (TM) data on 1:1 million scale. The length and direction of each lineament was measured and the lineaments were grouped based on length and direction. The nature of each lineament was assessed using the satellite data in conjunction with the collateral data. Remote sensing data due to its synoptic nature is found to be very useful in mapping lineaments. Images taken in the near infrared (NIR) region (0.7–1.1 μm) depicts clearly more lineaments than other bands. Radar data also provides information on lineaments due to its oblique look angles. Thermal infra-red (TIR) data is found to be useful in delineating wet lineaments with moisture/water. Lineaments present in the forest areas, soil covered areas are also clearly visible on images thus enabling us to delineate better structural features.

Normally, it is difficult to decide whether the mapped lineament is a fault or not, but if there is a clear displacement/offset then the lineament can be identified as a fault. Integration of the lineament map with the available structural and geological information of the terrain plus fieldwork helps to decide the nature of the lineament. Lineaments, which are seen as linear features on satellite data were mapped. Lineaments with length more than 100 km, i.e., major lineaments were mapped first from individual scenes of Landsat data on 1:1 million scale scenes; these were transferred to base map of the study area and make a single map. This map was superimposed on physical/road network map of the study area to eliminate any cultural lineament like road, railway lines which also appear as lineaments in the satellite image. IRS-1D WiFS FCC (Fig. 2.8) on 1:1 million scale was used further to map the lineaments. Lineaments were numbered and their length and direction were measured.

Based on the correlation between major lineaments and earthquake occurrence, the seismically active lineaments were identified. Studies by Ganesha Raj and Nijagunappa [101], Sitharam et al. [333, 334] were toward the identification of lineaments using satellite data for the state of Karnataka. Earthquakes with magnitude four and above that occurred in Karnataka and regions inside 300 km from the Karnataka state boundary were considered and overlaid on major lineament map of Karnataka and adjoining regions to assess the correlation between earthquakes and major lineaments. Currently, along many lineaments rivers/streams are flowing. Parts of the river courses of Hemavathi, Tunga, Yagachi, Vedavathi, Krishna, Nethravathi, Malaprabha, etc. follow major lineaments. The major lineaments in the study area, which have some correlation with the occurrence of earthquakes, were identified and shown in Fig. 2.8. The declustered earthquake data in unified moment magnitude scale have to be superimposed on the digitized map of extracting tectonic features. To characterize the seismic sources, the maximum reported magnitude and number of earthquake events associated with each of these sources are noted. There are some earthquake events which are not falling along any of the identified faults. In those cases the linear source model alone may not be able to give the correct picture of seismic hazard levels. To overcome this limitation three alternative source models can be adopted—point sources, smoothed gridded sources, and areal sources. These are widely accepted source models for evaluation of seismic hazard for regions where the sources are not clearly delineated.

2.5.2 Smoothed Point Source Model

The point sources are considered for hazard analysis using deterministic methods. The selection of point sources is done using two steps—discretizing the observed seismic activity and then smoothing the discretized seismic events. In this method, the study area is divided into small grids and the maximum earthquake magnitude in each grid is assigned to the center of that grid. The maximum earthquake magnitudes, thus obtained are smoothed using a smoothing window. The smoothing of the data is done to account for the source dimension of the earthquake events and for location errors [73, 255]. This type of source characterization was used

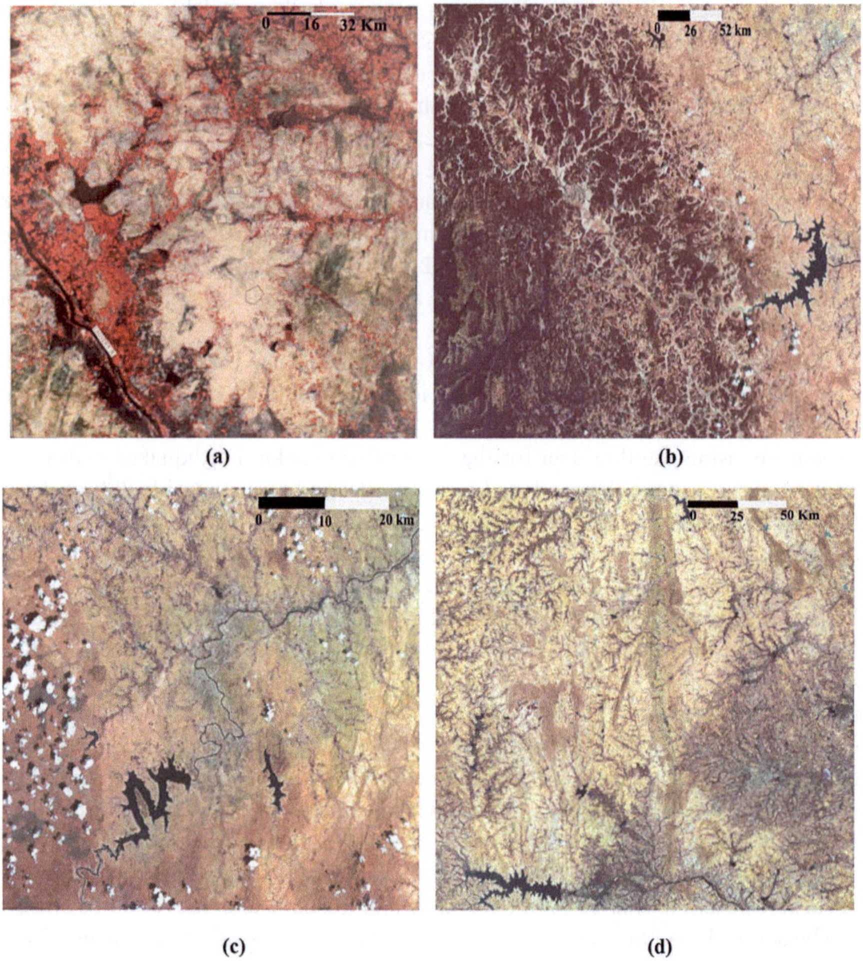

Fig. 2.8 IRS LISS-II images showing various lineaments in Karnataka (extracted from RS image procured from National Remote Sensing Agency in the year 2006 after [101]). (**a**) Cauvery lineament. (**b**) Hemavathy-Thirthahalli lineament. (**c**) Kabini lineament. (**d**) Arkavathi-Madhugiri lineament

for deterministic seismic hazard analysis (DSHA) by Costa et al. [73], Panza et al. [255], El-Sayed et al. [83]. For the deterministic seismic hazard evaluation of India, Parvez et al. [261], Kolathayar et al. [177] have modelled the seismic sources in India as point sources based on the same methodology.

2.5.3 Smoothed Gridded Seismicity Model

Gridded seismicity model [96, 386] is based on the seismic activity rate obtained from the earthquake catalogue and it is one of the most widely adopted methods to model seismic sources for the regions in the absence of clearly identified seismic sources. Some of the seismic hazard studies which considered zoneless approach for source identification are [149, 160, 189, 215, 373, 378]. In this method the study area is divided into grids and the number of earthquakes, which are having a magnitude higher than a cut-off magnitude, in each grid is counted. This will give the activity rate for that particular grid cell. Based on this value, the recurrence rates for different magnitude intervals can be calculated and these values can be smoothed using a Gaussian function (2.4) to get the final activity rate for each grid cell. The uncertainty involved in estimating the location of the earthquake event and the size of the seismic source can be accounted by this smoothing.

While considering these sources, the first step is the selection of grid size and the cut-off magnitude (M_{cut}). After selecting a suitable grid size, the number of earthquake events of magnitude greater than or equal to M_{cut} has to be calculated for each grid cell and this represents the maximum likelihood estimate of the total number of earthquakes (equal to or greater than M_{cut}) for that grid cell. In the present study the value of M_{cut} was taken as 4.0 to eliminate the effect of rock bursts, blasting. Moreover seismic events with magnitude less than 4.0 may not cause much damage also. Based on this value the recurrence rates for different magnitude intervals were calculated. These values were smoothed using a Gaussian function to get the final corrected values for each grid. This smoothing is done to account for the uncertainty associated with the location of earthquake events.

$$\hat{n}_i = \frac{\sum_j n_j e^{\frac{-\Delta_{ij}^2}{c^2}}}{\sum_j e^{\frac{-\Delta_{ij}^2}{c^2}}} \tag{2.4}$$

where n_j is the number of earthquakes in the jth grid cell and $\hat{n}_i$ is the smoothed number of earthquakes in ith cell; c is the correlation distance (to account for the location uncertainties) and Δ_{ij} is the distance between the ith and jth cells. In the present study a cut-off magnitude of $M_W = 4$ and a correlation distance of 50 km is used in smoothing the 'a' values. Since the linear source model was also considered in the present study, a single correlation distance was used.

2.5.4 Areal Source Model

For the hazard estimation using areal sources, the territory under study should be first divided into seismic sources such that, within a seismic source, earthquake-occurrence process is independent. For each seismic source, magnitude exceedance rates are estimated by means of statistical analysis of earthquake catalogs. These rates are the number of earthquakes per unit time, in which a specific magnitude is exceeded, and they represent the seismicity of the source [253].

The areal sources can be modelled as polygon-dipping areas which allow for local variations in seismicity characteristics for the other types of source models; linear sources and gridded seismicity approach (for example, changes in a, b-values, M_{max} etc.). In this method, a spatial integration process is carried out to account for all possible focal locations with an assumption that, within a seismic source, all points are equally likely to be an earthquake focus. Within a source, seismicity is evenly distributed by unit area and to correctly account for this modelling assumption, a spatial integration is performed by subdividing the original sources. Once subdivided into sub-sources, all the seismicity associated with a sub-source is assigned to a single point, and then the spatial integration adopts a summation form.

2.6 Seismicity Analysis

The seismic activity of a region is given by the Gutenberg–Richter earthquake recurrence law [112]. The recurrence rate given by this to this law is:

$$\log_{10} \lambda_M = a - bM \tag{2.5}$$

where λ_M is the total number of earthquakes with magnitude M and above which will occur in a year and a and b are the seismicity parameters of the region. The values of seismicity parameters can be evaluated using the maximum likelihood estimation technique [4, 369]. The parameter 'a' describes the productivity of volume, and b the slope of the frequency-magnitude distribution (FMD) describes the relative size distribution of events. The high-quality earthquake catalogs collected primarily in recent years and the availability of increased computing power have enabled researchers to investigate spatial variations in b with high precision. The strong difference in b is simply a reflection of the heterogeneity of the earth that emerges on all scales, once suitable datasets become available [382]. For evaluating the seismicity parameter, the completeness of the catalogue has to be analyzed and the data in the complete part of the catalogue need to be used for the analysis.

2.6.1 *Stepp's Method*

The completeness of the catalogue can be analyzed using the method suggested by Stepp [342] which will give the time interval in which the magnitude range is homogeneous. In the method proposed by Stepp [342], the sample mean is inversely proportional to the number of observations in the sample. The earthquake sequence is modelled based on a Poisson's distribution for obtaining an accurate estimate of the variance of the sample mean. Let $x_1, x_2, x_3, \ldots x_n$ be the number of events per unit time interval, then the unbiased estimate of the mean per unit interval for this sample is given by Eq. (2.6).

$$\lambda = \frac{1}{n}\sum_{i=1}^{n} X_i \tag{2.6}$$

where n is the unit time interval and its variance is $\sigma_\lambda^2 = \frac{\lambda}{n}$. When the time interval is taken as 1 year, the standard deviation of the above equation becomes $\sigma_\lambda = \frac{\sqrt{\lambda}}{\sqrt{T}}$; where T is the sample length. If this process is assumed to be stationary, it can be concluded that the standard deviation behaves as $\frac{1}{\sqrt{T}}$ in the subinterval of the complete years of the sample. Hence during this period the mean rate of occurrence in a magnitude class will be constant.

As a first step for the evaluation of the completeness period, the number of earthquakes reported during each decade for the given magnitude ranges is to be evaluated. The earthquake data is considered as complete as long as its variation is along the $\frac{1}{\sqrt{T}}$ line. The completeness periods for different ranges of magnitude can be estimated using this approach. The completeness period of the smaller magnitude ranges will be low as the globe was not well instrumented in the past. Hence lots of smaller magnitude events were not recorded properly. However the probability of events getting recorded increases with increase in earthquake magnitude. That is the reason for the increasing trend in completeness period with increase in magnitude. The seismicity parameters have to be evaluated based on the complete part of the catalogue.

2.6.2 *Magnitude of Completeness*

The magnitude of completeness is the lowest magnitude above which the earthquake recording is assumed to be complete. The magnitude of completeness (M_c) is defined as the lowest magnitude at which 100% of the events in a space-time volume are detected [294]. Below this magnitude a fraction of events is missed by the network because they are either too small to be recorded by enough stations or because they are below the magnitude of interest or because they are mixed with the coda of a larger event and therefore they passed undetected. There will be spatial and temporal variation of M_c and it will decrease with time mainly because of the increase in the number of seismographs in the region. The 'b' value can be evaluated using the maximum likelihood estimate as given below.

$$b = \frac{\log_{10}(e)}{\left[m_{\text{mean}} - \left(M_c - \frac{\Delta m_{\text{bin}}}{2}\right)\right]} \tag{2.7}$$

where m_{mean}—mean magnitude of the sample; M_c—magnitude of completeness and Δm_{bin}—magnitude bin size (incremental M_W values).

One of the most widely used methods to evaluate M_c is based on a power law fit for the frequency magnitude distribution (FMD), which was suggested by Wiemer and Wyss [381]. In this method a series of synthetic magnitude distributions are developed for each magnitude interval using maximum likely hood estimate. These '*a*' and '*b*' values obtained in the synthetic distribution are compared with the observed distribution and the goodness of fit is calculated. For calculating the goodness of fit, the absolute difference (R) between the observed and synthetic distribution has to be calculated as per Eq. (2.8).

$$R(a, b, M_i) = 100 - \left[\frac{\sum_{m_i}^{m_{\max}} (B_i - S_i)}{\sum_i B_i}\right] \tag{2.8}$$

where B_i and S_i are the observed and predicted cumulative number of events in each magnitude bin. The M_c value is then calculated at the magnitude where the goodness of fit is more than 90%. The goodness of fit can be tested for 95% also, but this level is rarely obtained for real earthquake catalogues [385]. The correct estimate of the *a* and *b* values depends critically on the completeness of the sample under investigation. The FMD deviates from a linear power law fit increasingly for smaller magnitudes which is caused by the fact that the recording network is only capable of recording a fraction of all events for magnitudes smaller than the magnitude of completeness, M_c. If M_c is raised to large values, the uncertainty in the *b* value estimate increases strongly. The situation is complicated by the fact that M_c varies as a function of space and time throughout all earthquake catalogs, hence estimating the correct M_c, while maximizing the available number of earthquakes, becomes difficult.

Figure 2.9 shows a typical FMD for Hindukush region (Source zone 30 by Kolathayar and Sitharam [175]) with *a b* value of 0.97, a value of 8.04, and Magnitude of completeness 4.7. The higher *a* value obtained implies that the average number of earthquakes per year in the region is very high. The *b* value nearing to one indicates that the size of earthquakes is evenly distributed in the region which means there is no large variation in the number of small magnitude events and large magnitude events. The magnitude of completeness of 4.7 implies that the earthquake catalog of the region is complete for all events with magnitude greater than 4.7.

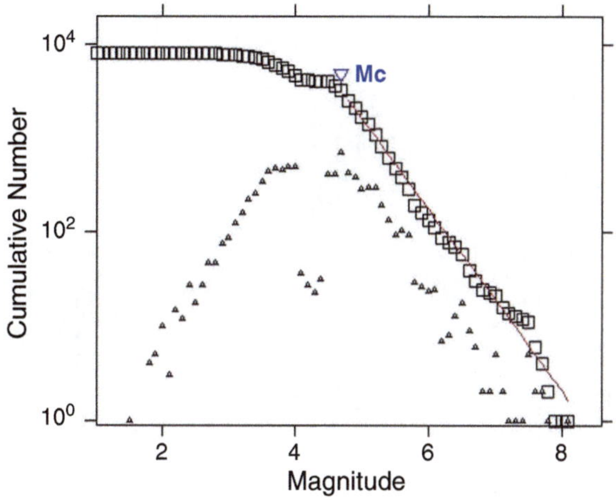

Fig. 2.9 The frequency magnitude distribution for Hindukush region (source: zone 30 of [175])

2.7 Preparation of Seismotectonic Atlas

The compilation and integration of all available data on geological, geophysical, and seismological attributes for the entire country is required for the proper evaluation of seismicity in different tectonic regions. In this context, well-defined and documented seismic sources published by authorized Geological agencies aid in identifying the potential seismic sources in a region. Maps are the representation of a geographic area on a piece of paper or canvas. These maps or Aerial photographs are never useful for computers. So in order to study the maps or aerial images it is necessary to convert them into digital form with appropriate resolutions to get high quality digital images.

The power of GIS technology makes creating and modifying maps a very simple task. Using colors and symbols, many different attributes can be highlighted. Application of GIS techniques allows inserting, extracting, handling, managing, and analyzing the data for the zoning of seismicity. The capability of Geographic Information Systems (GIS) to store and process data and images makes them very valuable in the field of seismic studies. The scanned maps can be georegistered and digitized using appropriate GIS tools. The coordinates are to be fixed based on the latitudes and longitudes of the region in the map.

2.7.1 Earthquake Catalog for India

Kolathayar and Sitharam [175] created an updated earthquake catalog for India and adjoining regions. They collected the details of earthquake events for the period from 250 BC to 1505 AD from [79]. The later portion of historic earthquakes was compiled from the work of various researchers [21, 27, 30, 62, 63, 109, 159, 170,

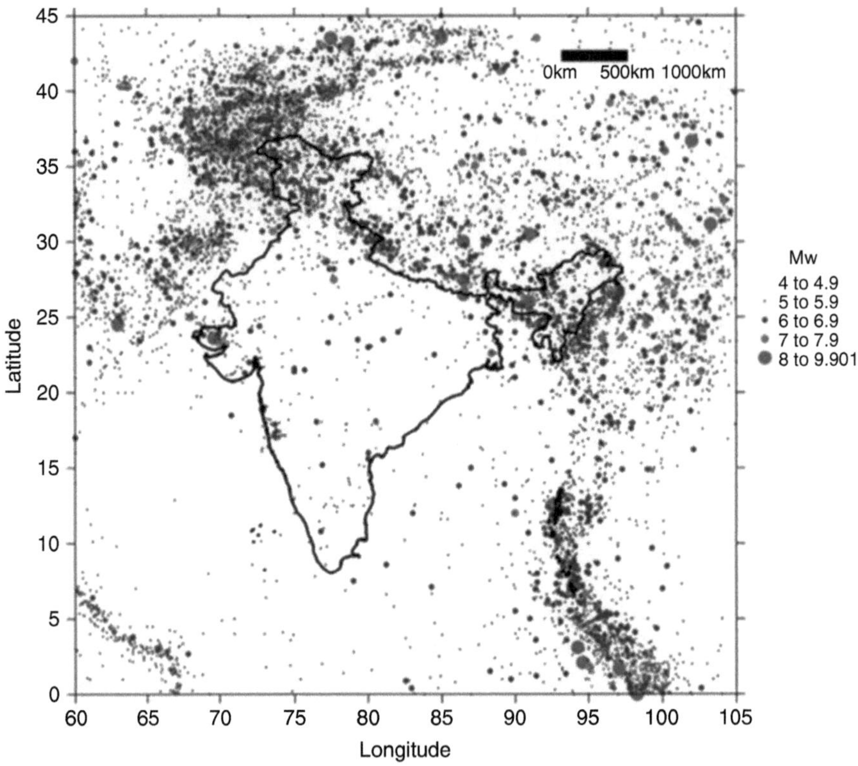

Fig. 2.10 Distribution of earthquake events in and around India (after [175])

250, 278, 280, 341, 351]. They compiled the instrumental catalog from national and international agencies like Indian Meteorological Department (IMD), National Geophysical Research Institute (NGRI) Hyderabad, International Seismological Center (ISC), data file (for the time period between 1964 and 2010), Harvard seismology and USGS/NEIC catalog (for the time period between 1973 and 2010). They have developed region specific magnitude conversion relations to homogenize the entire catalog in unified moment magnitude scale. Further, the catalog was declustered to remove aftershocks and foreshocks. The distribution of earthquake events of magnitude greater than 4, in and around India is presented in Fig. 2.10.

2.7.2 Development of Seismotectonic Map for India

Geological Survey of India has compiled all the available geological, geophysical, and seismological data for entire India. Seismotectonic [316] atlas contains 43 maps covering entire India and adjoining areas, with all available data related to

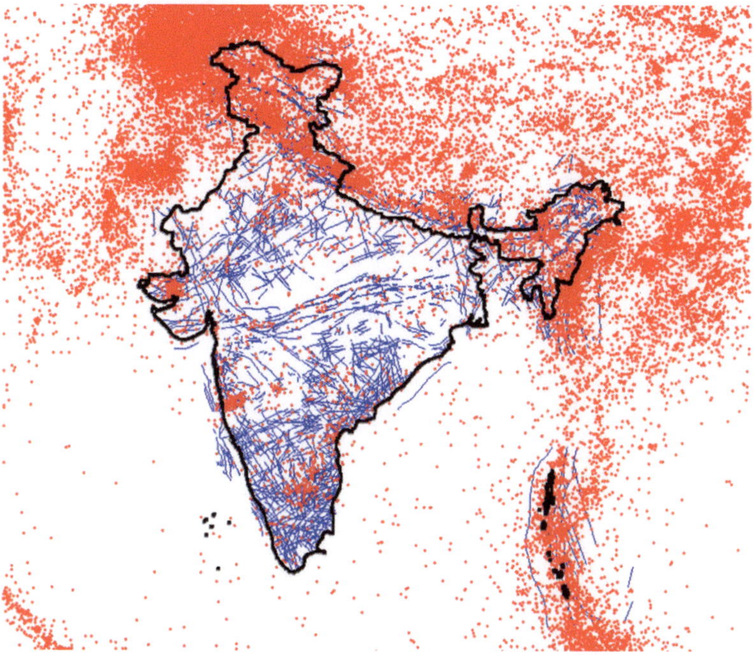

Fig. 2.11 Seismotectonic map of India showing linear sources and events of $M_W \geq 4$ (after [175])

earthquakes. It is a compilation of multi-thematic database comprising of 43 maps (presented in 42 sheets) covering India and adjacent regions of neighboring countries on 1:1 million scale. Kolathayar and Sitharam [175] developed seismotectonic map for India where all the 42 sheets in SEISAT published by GSI were scanned and the faults, shear zones, lineaments, and other seismic sources were digitized. Then the complete homogenized earthquake events [175] were superimposed on this source map to get the final digital seismotectonic map of India with all the seismic sources and earthquake events (Fig. 2.11).

Chapter 3
Seismic Hazard Analysis

3.1 Introduction

The earthquake resistant design aims to prepare the structure to withstand a certain level of ground shaking without undergoing extensive damages. Hence a reliable estimation of ground motion parameter at a given site forms a pre-requisite for any earthquake resistant design procedure. Seismic hazard analysis involves the quantitative estimation of ground shaking hazard at a particular site. This chapter describes the estimation of ground motion parameters such as peak horizontal acceleration (PHA) and spectral acceleration (S_a) for the state of Karnataka and Kalpakkam nuclear power plant (NPP) site.

3.2 Ground Motion Prediction Equations

Ground motion amplitude values will vary with variation of distance from the earthquake source. The ground motion prediction equations (GMPE or attenuation relations) are being used to predict the attenuation properties in any region. The most common parameters which are used to measure the ground motion amplitude are the peak horizontal acceleration (PHA) and peak horizontal velocity (PHV). The PHA values are more commonly used to describe the earthquake ground motion because of their direct association with the inertial forces [181]. It has been found that the PHV is a good parameter for characterization of ground motion with intermediate frequencies for structures like tall or flexible buildings, bridges, etc. The ground motion prediction equation gives the variation of ground motion parameter as a function of earthquake magnitudes and the source-to-site distance. It is typically of the form as shown in Eq. (3.1).

$$\ln y = c_1 + c_2(M - 6) + c_3(M - 6)^2 - \ln(R) - c_4(R) + \ln(\varepsilon) \tag{3.1}$$

T. G. Sitharam et al., *Comprehensive Seismic Zonation Schemes for Regions at Different Scales*, https://doi.org/10.1007/978-3-319-89659-5_3

where y, M, R, and ε refer to PGA, spectral acceleration (g), moment magnitude, hypocentral distance, and error associated with the regression, respectively.

The GMPE are broadly classified into two categories, viz. for inter plate earthquake regions and for mid-plate or stable continental shield regions. This is necessitated by the fact that the dissipation of seismic energy occurs at a faster rate at plate boundaries than at the mid plate regions. Some of the relations developed for active tectonic regions (inter plate regions) are [37, 54, 55, 156, 269, 296, 318]. Some of the attenuation relations for the stable continental shield regions are by Toro et al. [359], Atkinson and Boore [16] and Raghu Kanth and Iyengar [274]. Table 3.1 lists the popular ground motion prediction equations developed worldwide in recent times.

In India, there is a lack of strong motion data and this in turn has resulted in the development of only very few region specific GMPEs. Some of the important GMPE available in India are [318] for the Himalayan region, [146] for Delhi region; [274] for Peninsular India; [234] for Sikkim Himalaya; [236] for Guwahati and [322] for Himalayan Region. Out of these attenuation relations, the most widely used relations are [274] and [322].

Table 3.1 List of various attenuation relations proposed by various authors for different regions

Author	Region
Iyengar and Ghosh [146]	Himalayan region
Campbell and Bozorgnia [55]	ENA
Yu and Wang [401]	NE Tibetan Plateau region
Atkinson and Boore [16]	ENA
Raghu Kanth and Iyengar [274]	Southern Indian Peninsula
Malagnini et al. [205]	San Francisco
Chiou and Youngs [68]	Japan, Mexico, California
Castro et al. [60]	Mexico
Ford et al. [94]	Northern California
Bennington et al. [23]	Parkfield, California
Nath et al. [235]	Garhwal Himalaya
Boore and Atkinson [37]	Worldwide
Sharma et al. [321]	Kaahchh region, Gujarat
Chun and Henderson [70]	North Korea
Sharma et al. [322]	Himalayan region
Nath et al. [236]	Guwahati
Ghasemi et al. [106]	Iran
Koulakov et al. [180]	Turkey
Gupta [110]	Indo-Myanmar subduction zone

3.3 Deterministic Seismic Hazard Analysis

The deterministic approach considers a particular earthquake scenario, either realistic or assumed one. The DSHA approach uses known seismic sources which are near the site and available historical seismic and geological data to generate discrete, single-valued events or models of ground motion at the site. The earthquakes are assumed to occur on the source closest to the site. The deterministic seismic hazard analysis needs three input details like earthquake source, controlling earthquake at the source, and an attenuation relation to evaluate the seismic hazard. A schematic diagram of different steps involved in DSHA methodology is given in Fig. 3.1. In DSHA, the controlling earthquake is assumed to act along the source at the shortest distance from the site. The uncertainties involved in the earthquake magnitude or location are not taken into account and this method will give an upper bound values for the ground motion. Hence DSHA method is adopted in evaluation of seismic hazard for some of the critical structures like nuclear power plants, big dams, bridges, hazardous waste contaminant facilities, etc. The results obtained from deterministic analysis can be used as a cap for the probabilistic analysis.

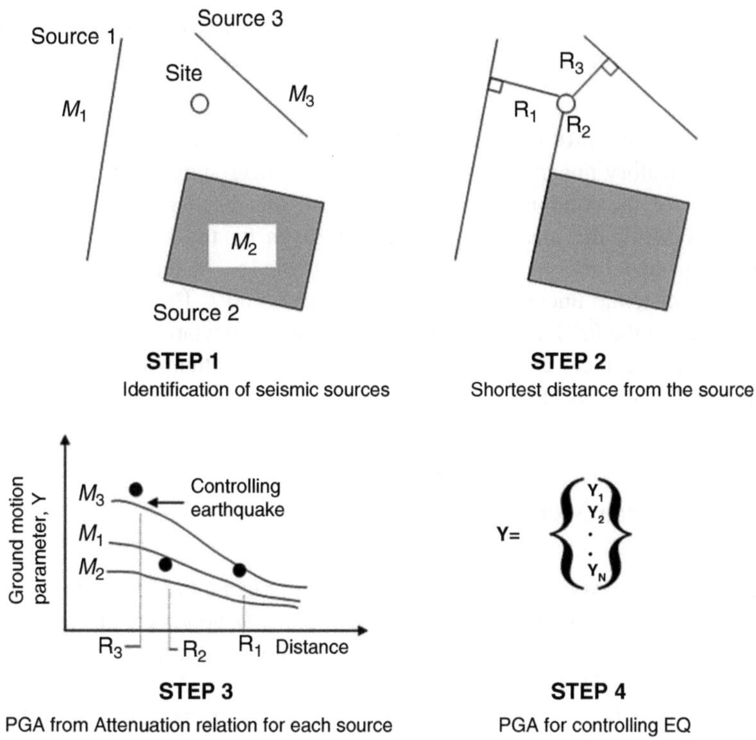

Fig. 3.1 Various steps involved in DSHA method (after [181])

3.4 Probabilistic Seismic Hazard Analysis

The evaluation of seismic hazard involves quantification of uncertainties in earthquake magnitude, location, recurrence rate, and the attenuation characteristics of seismic waves. The adoption of probabilistic approaches in seismic hazard analysis will provide a framework for identifying, quantifying, and combining these uncertainties in a rational way [181]. Probabilistic seismic hazard analysis (PSHA) was initially developed by Cornell [72]. Many researchers have adopted this methodology for evaluating hazard and recently this method has been adopted in India by Iyengar and Ghosh [146], Raghu Kanth and Iyengar [273], Anbazhagan et al. [10], Vipin et al. [376], Kolathayar and Sitharam [176] and Sitharam et al. [335] for the probabilistic seismic hazard analysis of Delhi, Mumbai, Bangalore, North West Himalaya regions, Peninsular India, Andaman & Nicobar regions, and India, respectively. The important steps involved in PSHA are shown in Fig. 3.2. The main steps involved in PSHA methods are

- Identification and characterization of earthquake sources
- Characterizing the earthquake recurrence rate
- Evaluation of ground motion using the attenuation relationships
- Determination of mean annual rate of exceedance of ground motion parameter by considering the uncertainties in earthquake location, size, and attenuation relation.

Two types of variability can be considered in PSHA studies—aleatory and epistemic. The aleatory uncertainty is because of the uncertainty in the data used and it accounts for the randomness associated with the result given by a particular model. In GMPE, the aleatory variability is given by the standard deviation of the mean ground motion. The incomplete knowledge in the predictive models causes the epistemic uncertainty (modelling uncertainty). The aleatory variability is considered in the PSHA by considering the standard deviation of the model error and the epistemic uncertainty is considered by multiple attenuation models for the evaluation of seismic hazard.

3.5 Logic Tree Methodology

The uncertainties involved in different models may make the selection of seismic hazard models difficult. In these cases the logic tree approach allows a formal characterization of epistemic uncertainty by including alternative models in the analysis [31, 52, 343]. Logic tree consists of a series of nodes and several models (hypothesis) can be assigned to each node as different branches. A subjective weightage can be given to each of these branches depending on the likelihood of being correct and the weightage for all the branches at a particular node should be equal to unity. In PSHA the different models are incorporated for magnitude

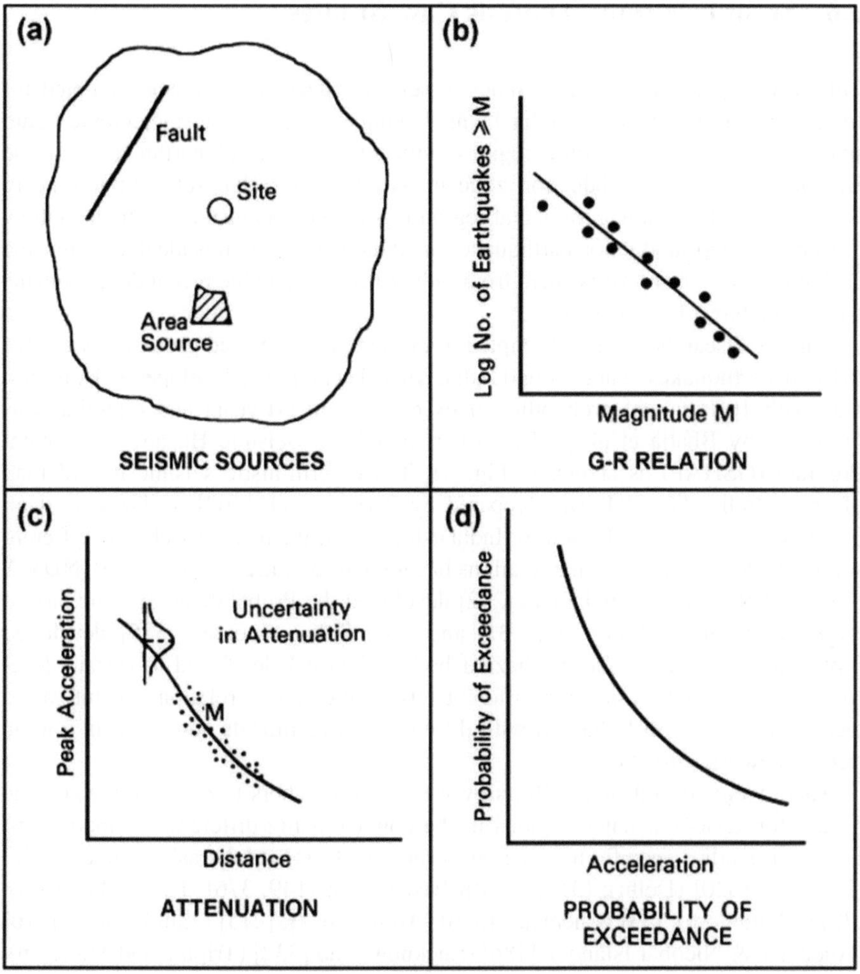

Fig. 3.2 Various steps involved in PSHA method (after [181])

recurrence rate, attenuation relation, estimation of source parameters, evaluation of maximum probable earthquake, etc. The weightage of the terminal branch can be obtained by multiplying the weightage of all the branches leading to it. Even though it is possible to include as many branches as [52] in the logic tree, this will increase the computational effort. Now almost all the PSHA studies are using a logic tree approach to capture the epistemic uncertainties.

3.6 Seismic Hazard Analysis Case Studies

This section presents the case studies of seismic hazard assessment attempted for India and adjoining areas. India being having a complex seismotectonic setup spatially with active tectonic regions with shallow crustal earthquakes in the Himalayas, Interface subduction zone in Kashmir, Intraslab subduction zone in North East India, stable continental region in southern peninsula and both shallow crustal and deep subduction earthquakes in Andaman region, it is ideal to review the case studies of hazard assessment from India to understand the recent developments on seismic hazard assessment.

Various researchers have attempted to evaluate the expected ground motion due to future earthquakes in and around India. Khattri et al. [171] developed PGA hazard map with 10% annual probability of exceedance in 50 years and a similar was presented by Bhatia et al. [25] as a part of Global Seismic Hazard Assessment Program (GSHAP) as shown in Fig. 3.3. The deterministic seismic hazard map of entire India (Fig. 3.4) was prepared by Parvez et al. [260]. Kolathayar et al. [177] estimated seismic hazard for India using deterministic approach with different source models and attenuation relations in a logic tree framework (Fig. 3.5). NDMA [237] and Nath and Thingbaijam [232] developed the Probabilistic Seismic hazard map for Indian landmass (Fig. 3.6 and 3.7). Sitharam et al. [335] developed probabilistic maps for seismic hazard both at bedrock level and at surface level for the Indian land mass using different sets of attenuation relations to suit varied tectonic provinces in India and with different source models combined in a logic tree framework (Fig. 3.8).

There were several other efforts by various researchers to estimate the seismic hazard for various isolated regions in the country using different methodologies. Few such studies specific to a region or city are [304] (Maharashtra state); [222] (Kolkata); [320] (Delhi); [319] (North East India); [149, 376] (Peninsular India); [327] (Bangalore); [33] (Chennai); [326] (Ahmedabad); [215] (Tamil Nadu); [176] (Andaman & Nicobar Islands), [186] (Lucknow) and [332] (Tripura and Mizoram).

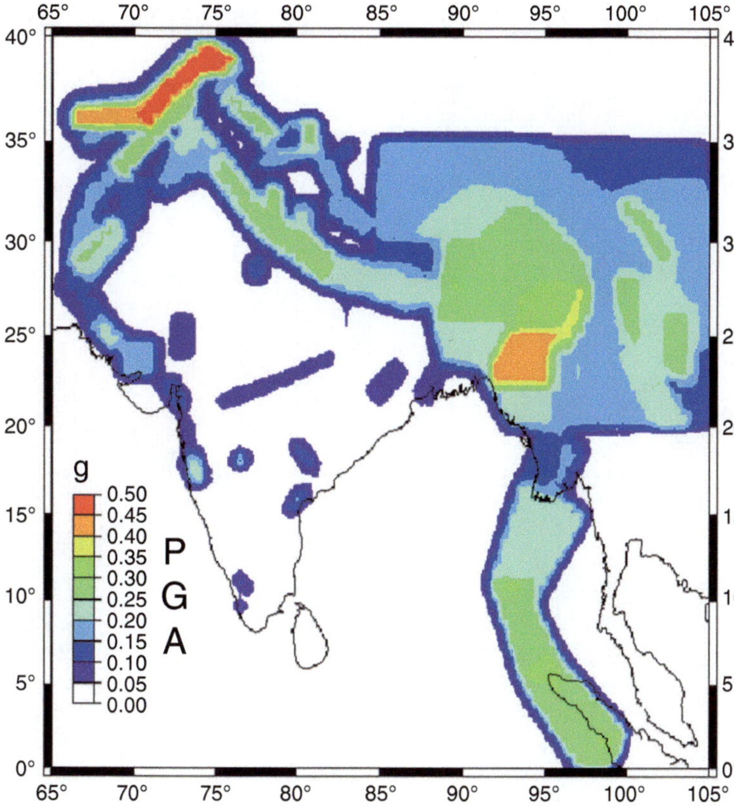

Fig. 3.3 Spatial variation of peak ground acceleration corresponding to 10% probability of occurrence after [25]

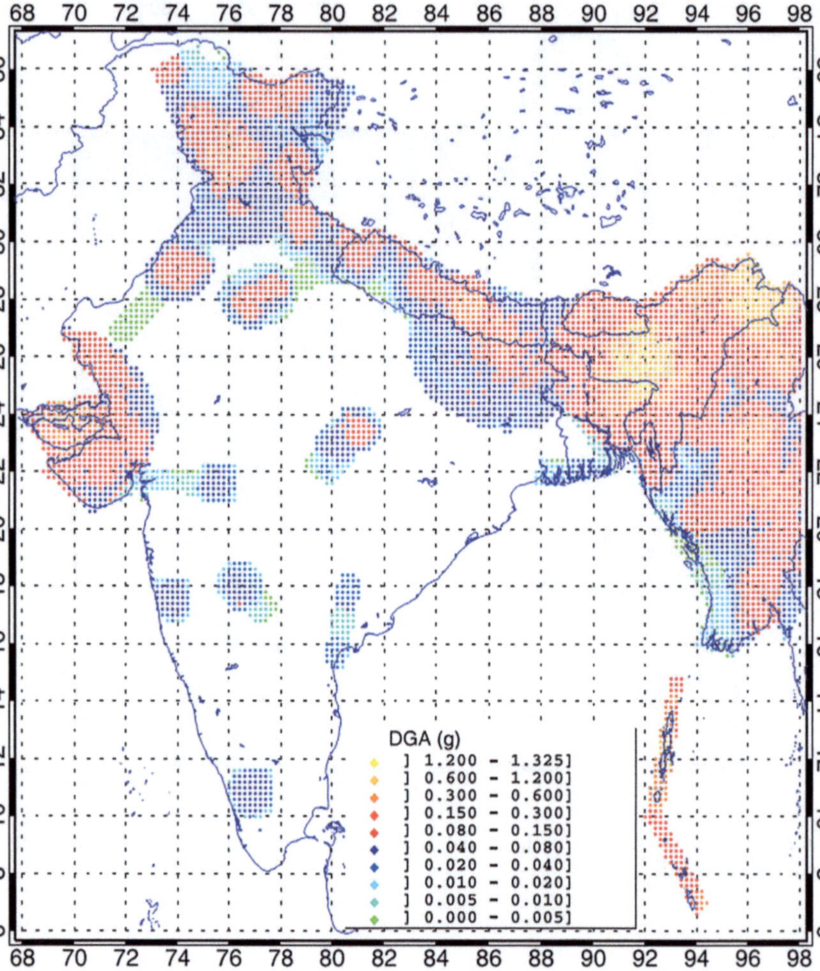

Fig. 3.4 Spatial distribution of the design ground acceleration after [260]

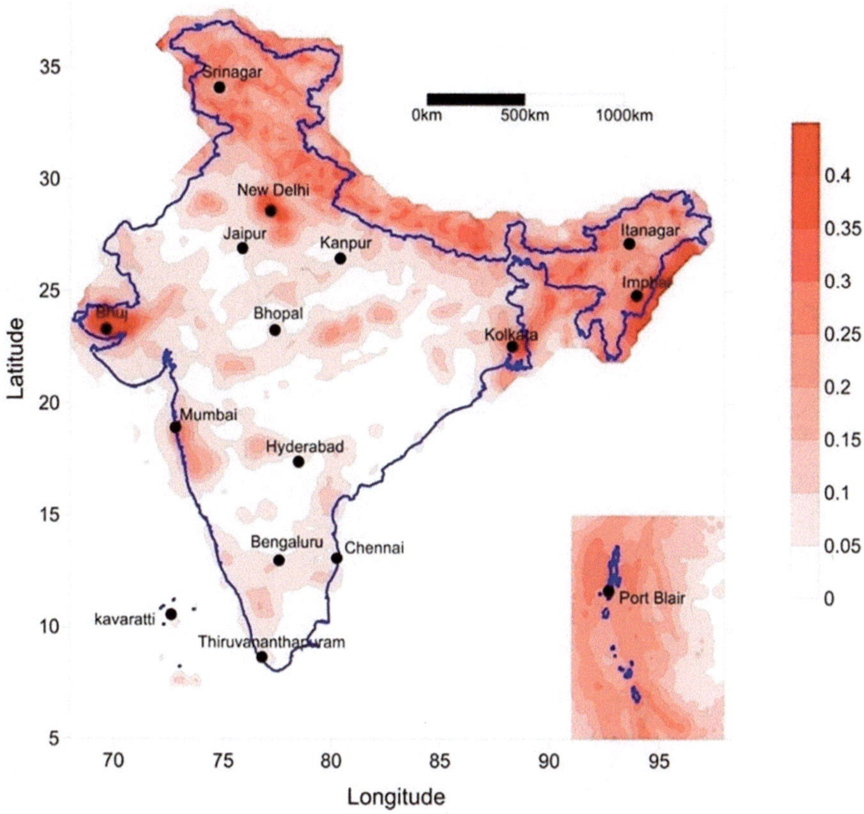

Fig. 3.5 Spatial distribution of the design ground acceleration after [177]

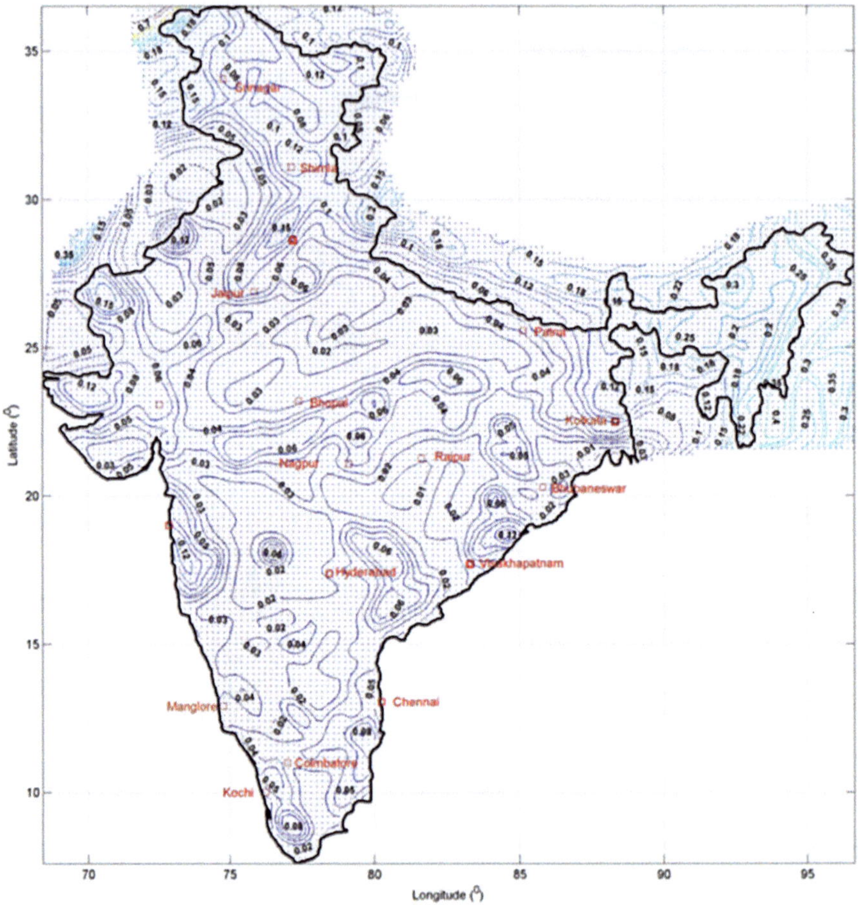

Fig. 3.6 PGA Contours with 10% probability of exceedance in 50 years (return period 500 years) on A-type sites [237]

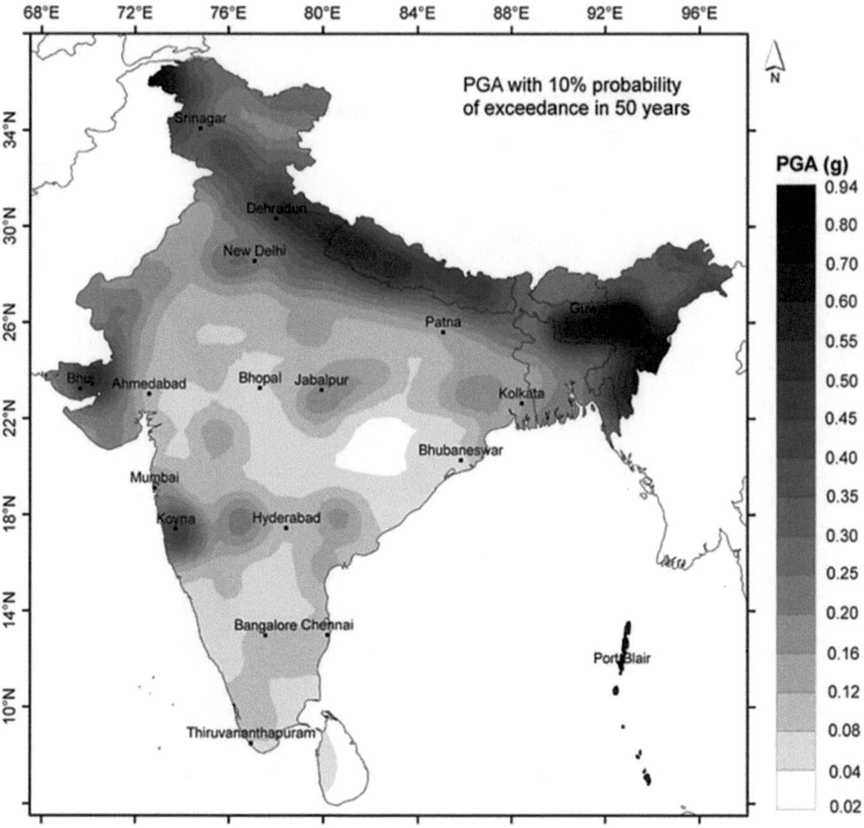

Fig. 3.7 Seismic hazard distribution in India in terms of PGA [232]

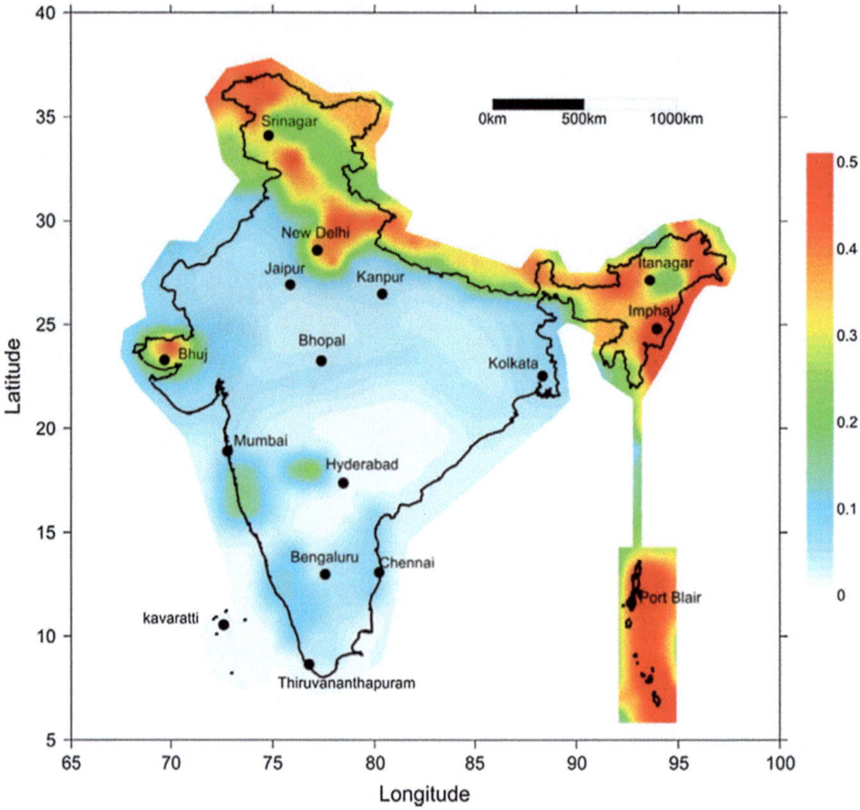

Fig. 3.8 PHA values (g) corresponding to a return period of 475 years (10% probability of exceedance in 50 years, PSHA) [335]

Chapter 4
Seismic Site Characterization

4.1 Introduction

Evidences from past earthquakes clearly show that the damages due to an earthquake and its severity are controlled mainly by three important factors i.e., earthquake source and path characteristics, local geological and geotechnical characteristics, structural design and construction features of structures. Seismic ground response at a site is strongly influenced by local geological and soil conditions. The exact details of the geological, geomorphological, and geotechnical data along with seismotectonic background and seismicity are needed to evaluate the ground response and site effects. The damage pattern during an earthquake depends on the soil characteristics at a site and it may have a major effect on the level of ground shaking. This point highlights the importance of site characterization in microzonation studies. The regional tectonic maps as well as surface geology maps and vertical geological profiles would be the essential ingredients for the seismic microzonation study. The characteristics and thickness of site's soil conditions are to be identified based on borings, in-situ geophysical and geotechnical tests. The geological, geomorphological, and geotechnical databases are needed for assessing the local site effects for site amplification as well as for liquefaction and landslide susceptibility.

The geometry of the subsoil structure, the soil types and the variation of their properties with depth, the lateral discontinuities and the surface topography influence the amplification of ground motion and hence of intensive damage during destructive earthquakes. For this the accurate knowledge of the geology, geomorphology, geophysical data, and geotechnical details are the key parameters controlling the damage severity during an earthquake. Thus the seismic site characterization has its significance in seismic microzonation studies. The present chapter describes in detail about the different methodologies for site characterization.

T. G. Sitharam et al., *Comprehensive Seismic Zonation Schemes for Regions at Different Scales*, https://doi.org/10.1007/978-3-319-89659-5_4

Further, this chapter also demonstrates the application of suitable methodology to perform seismic site characterization for two study areas both at micro and macro-level.

4.2 Methods for Seismic Site Characterization

Local geological conditions have a significant effect on the earthquake ground motion at a given site. The main aim of the site characterization study is to evaluate basic and engineering properties of soil and then categorize the soil stratum based on these soil properties. Seismic site characterization is helpful for assessing various hazards associated with the earthquake at the ground surface level. It involves site investigation, data collation, interpretation, categorizing into various soil classes and representing the spatial distribution. The site characterization methods can be broadly divided into the following categories as (1) based on the local geology (2) based on the geotechnical field test data (3) based on the geophysical test data (4) based on the topographic slope.

4.2.1 Based on Geology

The surface geology is an excellent parameter for carrying out first level site characterization. It can be correlated to the different site classes given by various codes like NEHRP [51], Eurocode [86]. Wills et al. [384] has refined the shear wave velocity classification of various geologic units and prepared a site condition map for the California. They also categorized various geological units in NEHRP site classes. The site class B consists of plutonic and metamorphic rocks, most volcanic rocks, coarse sedimentary rocks of Cretaceous age and older. The Franciscan Complex rocks other than melange and serpentine, crystalline rocks of the transverse ranges which tend to be more sheared, Cretaceous siltstones, or mudstone, etc. comes under class BC. The site class C consists of Franciscan melange and serpentine, sedimentary rocks of Oligocene to Cretaceous age, or coarse-grained and younger sedimentary rocks.

The sedimentary rocks of Miocene and younger age, unless formation is notably coarse grained, Plio-Pleistocene alluvial units, older (Pleistocene) alluvium, some areas of coarse younger alluvium are categorized in class CD. The site class D mainly consist of younger (Holocene) alluvium. The fill over bay mud in the San Francisco Bay area, fine grained alluvial, and estuarine deposits elsewhere along the coast come under DE. The site class E consists of bay mud and similar intertidal mud. Stewart et al. [344] has classified a total of 427 California recording stations based on mapped surface geology. Table 4.1 presents the classification of each site into three geologic schemes based on geologic age, depositional environment, and material texture. The site characterization based on local geology is an approximate

Table 4.1 Criteria for surface geology classifications [344]

Age	Depositional environment	Sediment texture
Holocene	Fan alluvium	Coarse
Pleistocene	Valley alluvium	Fine
	Lacustrine/marine	Mixed
	Aeolian	
	Artificial Fill	
Tertiary		
Mesozoic + Igneous		

method and very good for a first level site characterization. However there are lot of issues associated with this method such as (1) it does not address the variability of V_s within a particular geological unit and (2) it does not provide any information regarding the thickness of a particular unit which also determines the variability of V_{s30} [379].

4.2.2 Based on the Geotechnical Field Test Data

Site response and ground failure are strongly influenced by the properties of soil. Site exploration usually begins with a thorough review of the available information about the site and its surroundings. Geotechnical reports for the sites may be available from various governmental or nongovernmental agencies. Geotechnical site characterization requires a full 3D representation of stratigraphy, estimates of geotechnical parameters and hydrogeological conditions and properties. The traditional methods like drilling and undisturbed sampling can provide adequate stratigraphic details and estimates of geotechnical parameters. But they cannot provide useful estimates of hydrogeological conditions which are very important in estimating the seismic hazards like liquefaction, landslides, etc. Site response is primarily influenced by the properties that influence wave propagation, particularly stiffness and damping. Ground failure is influenced by the shear strength of the soil. Soils are highly non-linear even at very low strains. This non-linearity causes soil stiffness to decrease and damping to increase with increasing strain amplitude. Both site response and ground failure are parts of the same continuous spectrum of non-linear soil behavior.

The in-situ tests generally conducted to identify the soil stratification and engineering properties of the soil layers are penetration tests. Two methods that have been widely used are the Standard Penetration Test (SPT) and Cone Penetration Test (CPT). Standard Penetration Test is generally used to investigate cohesionless or relatively stiff soil deposits, whereas CPT is used to identify soil properties in soft soil deposits [200]. SPT is generally performed in a borehole and measures soil resistance in terms of the number of blows required for 30 cm penetration of standard sampler. The blow for the penetration is given by a 63.5 kg hammer, free falling from 75 cm height. The variability of the Standard Penetration Test equipment and procedures used has significant effects on the obtained blow counts [315, 336]. The energy delivered to the split-spoon sampler is strongly influenced by many factors such as hammer type, borehole diameter, rod length, rod diameter, tightness of the rod joints, verticality of the rod string, and type of sampler, etc. Therefore it is very important to have sufficient information to estimate the energy ratio correction for SPT blow counts before using these results for assessing the properties of soil layers. SPT is the most fundamental field test and is generally carried out as a part of subsurface investigation, prior to construction. Hence there are many correlations available relating different soil parameters to the SPT-N values. The number of blows obtained from the SPT is then subjected to corrections for overburden as well as dilatancy [29]. BIS-1893 [28] has classified the soil into three major categories: Type I, II, and III based on the SPT-N value range. Table 4.2 presents a site classification proposed by BIS-1893 [28].

Empirical relations have been proposed to correlate the penetration test results between CPT and SPT [289] as well as with the shear-wave velocities [147, 213, 245].

James et al. [153] has developed a similar correlation between shear wave velocity and SPT-N value is developed for the east coast region of South India. The N-values measured in the field were subjected to various corrections such as (1) Overburden pressure correction (C_N), (2) Hammer energy correction (C_E), (3) Borehole diameter correction (C_B), (4) liner correction (C_S), (5) Rod length correction (C_R), and (6) correction for fine contents (C_{FC}) [61, 313, 336, 400]. The SPT value corrected for 60% energy efficiency and 100 kPa overburden pressure $(N_1)_{60}$ was evaluated using Eq. (4.1) [287]. The $(N_1)_{60cs}$ provides consistent value for penetration resistance and hence become a standard for assessing liquefaction potential [291]. They have also compared V_s-SPT correlation with the correlation developed by Anbazhagan and Sitharam [9] as in Fig. 4.2.

$$(N_1)_{60} = N \times [C_N \times C_E \times C_B \times C_S \times C_R] \tag{4.1}$$

Table 4.2 Soil classification based on SPT-N values [28]

Soil type	Description	SPT-N value range
Type I: rock or hard soil	Well graded gravel and sand gravel mixtures with or without clay binder, and clayey sands poorly graded or sand clay mixtures (GB, CW, SB, SW, and SC)	>30
Type II: medium soils	All other soils like poorly graded sands or gravelly sands with little or no fines (SP)	10–30
Type III: soft soils	All soils other than SP	<10

The obtained $(N_1)_{60}$ was again corrected for fine content as suggested by Idriss and Boulanger [131] using Eqs. (4.2) and (4.3).

$$(N_1)_{60cs} = (N_1)_{60} + \Delta(N_1)_{60} \tag{4.2}$$

where

$$\Delta(N_1)_{60} = \exp\left[1.63 + \left(\frac{9.7}{FC + 0.1}\right) - \left(\frac{15.7}{FC + 0.1}\right)^2\right] \tag{4.3}$$

Here FC is the fine content, i.e. percentage of dry weight finer than 75 µm. The V_s values estimated from the MASW results are correlated to corrected SPT N values at respective depths. The regression equation developed between shear wave velocity V_s and corrected SPT value $(N_1)_{60}$ is having a regression coefficient of 0.63. Figure 4.1 shows the correlation between V_s and $(N_1)_{60}$ for the study area. Figure 4.2 presents a comparison between the correlation proposed by Anbazhagan & Sitharam [9] and James et al. [153].

A list of some of the relationships proposed to calculate shear wave velocity in terms of SPT 'N' value is given in Table 4.3.

Cone penetration tests (CPT) are widely used particularly for soft clays, silts, and in fine to medium sand deposits [48]. The test consists of measuring the resistance to penetration of a standard cone into the ground at a rate of 10–20 mm/s. The total resistance against penetration has two components, (1) cone side resistance and (2) cone tip resistance. A comparison of the two methods shows that the CPT method

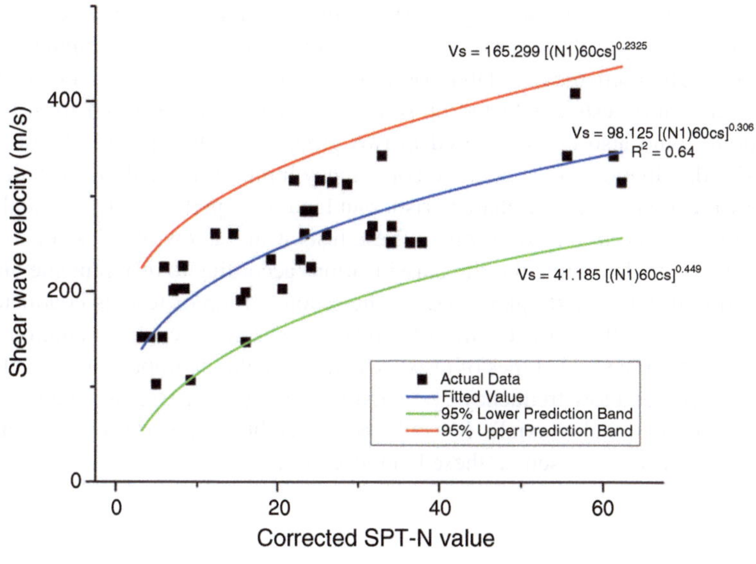

Fig. 4.1 Shear wave velocity vs corrected SPT-N value (after [153])

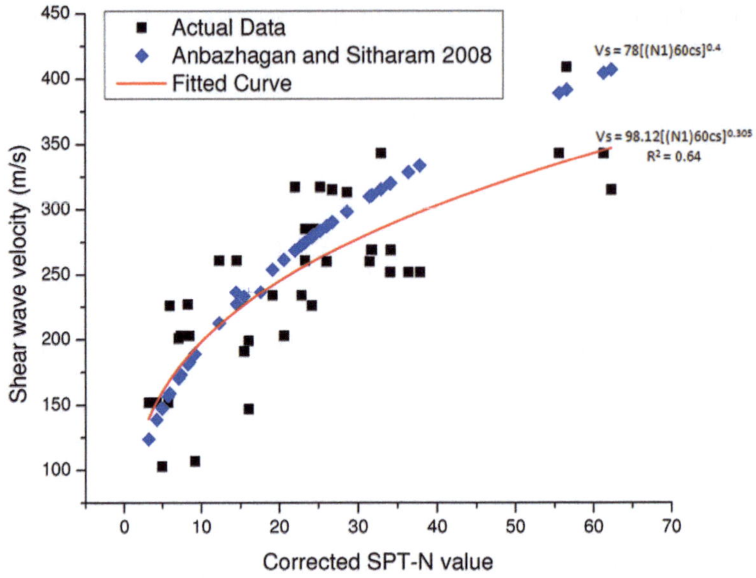

Fig. 4.2 Comparison between the correlation relations developed for corrected SPT-N values (after [153])

is advantageous over the SPT. The CPT gives a continuous profile of end resistance and side friction while the SPT only provides discrete values of blow count. The CPT also has better control over errors when compared with SPT. CPT offers a minimum disturbance to the underlying strata and hence is more suitable to site with underlying sensitive soil layers. However, the CPT is not suitable for strata containing dense gravels, soft rock as it is difficult to push the cone through this. The site classification schemes available based on the CPT values are given in Table 4.4.

The laboratory tests conducted on soil and rock samples retrieved from boring operations could also be considered in two groups. The first group of tests (i.e., grain size distribution, water content, consistency limits) is needed to determine the soil classification, grain size characteristics and index properties of the soil and rock layers encountered in the soil profile. These tests would allow the classification of the soil layers to determine the site classifications according to different site classes proposed in different earthquake codes. The second group of tests is conducted to obtain shear strength characteristics of soil specimens under cyclic excitations [11, 139, 174]. The three basic types of tests are resonant column, impulse wave velocity measurements, and low frequency cyclic loading tests (cyclic triaxial, cyclic simple shear, cyclic torsional triaxial). It would be preferable to determine the dynamic shear modulus curves based on these laboratory tests.

Table 4.3 Proposed relationships to estimate shear wave velocity from SPT value (after [238])

Author	Data	Soil type	V_s (m/s)
Kanai et al. [163]	Not known	All soils	$V_s = 19\,N^{0.6}$
Shibata [323]	Not known	Sandy soils	$V_s = 32\,N^{0.5}$
Imai and Yoshimura [135]	Not known	All soils	$V_s = 76\,N^{0.33}$
Ohba and Toriumi [242]	Not known	Alluvial soils	$V_s = 84\,N^{0.31}$
Ohta and Goto [245]	Not known	Sandy soils	$V_s = 87\,N^{0.36}$
Ohsaki and Iwasaki [244]	Not known	All soils Cohesionless soils	$V_s = 82\,N^{0.39}$ $V_s = 59\,N^{0.47}$
Imai and Yoshimura [136]	Not known	All soils	$V_s = 90\,N^{0.341}$
Imai [133]	Not known	All soils	$V_s = aN^b$ $a = 102 b = 0.29$(H.clay) $a = 81\ b = 0.33$(H.sand) $a = 114\ b = 0.29$(P.clay) $a = 97\ b = 0.32$(P.sand)
Ohta and Goto [245]	Not known	All soils	$V_s = 69\,N^{0.17}\,D^{0.2}EF$ $E = 1$(H); 1.3(P) $F = 1$ (clay); 1.09(f.sand); 1.07(m.sand); 1.14(c.sand); 1.14(c.sand); 1.15(g.sand); 1.45(gravel)
Seed and Idriss [310]	Not known	All soils	$V_s = 61\,N^{0.5}$
Imai and Tonouchi [134]	1654 sets of data (Japan)	All soils	$V_s = 96.9\,N^{0.314}$
Seed et al. [313]	Unknown	Sands	$V_s = 56.4\,N^{0.5}$
Sykora and Stokoe [349]	229 sets of crosshole data	Granular soils	$V_s = 106.7\,N^{0.27}$
Fumal and Tinsley [97]	Not known	Sands and gravelly sand Soils	$V_s = 152 + 5.1\,N^{0.27}$
Sykora and Koester [350]	186 points (7 m, 7 sites)	Holocene gravels	$V_s = 63\,N^{0.43}$
	186 points (7 m, 7 sites)	Pleistocene gravels	$V_s = 132\,N^{0.32}$
Okamoto et al. [248]	Not known	All	$V_s = 125\,N^{0.3}$ (P.sand)
Lee (1990)	Not known	Clays Sands Silts	$V_s = 114\,N^{0.31}$ $V_s = 57\,N^{0.49}$ $V_s = 106\,N^{0.32}$
Jafari et al. [148]	Not known	All soils	$V_s = 22\,N^{0.85}$
LIQUFAC	Unknown	All	$V_s = 243.8\sigma_e^{0.4}$
LIQUFAC	Unknown	All	$V_s = 152\sigma_e^{0.3}$
Anbazhagan and Sitharam [9]	SPT and MASW data	Sand	$V_s = 57(N_1)_{60,cs}^{0.44}$
		Clays	$V_s = 57\,N^{0.44}$

Where, N: SPT value; ε_e: effective vertical stress of the soils; V_s: shear wave velocity; D: depth (m); H: Holocene; P: Pleistocene; f: fine; m: medium; c: coarse; g: gravel

Table 4.4 Soil classification based on CPT N values [200]

Soil type	Cone penetration resistance q_c (MPa)
Soft clay and mud	<1
Moderately compact clay	1–5
Silt and loose sand	≤5
Compact to stiff clay and compact silt	>5
Moderately compact sand and gravel	5–12
Compact to very compact gravel	>12

4.2.3 Based on the Geophysical Test Data

Seismic tests are classified into borehole (invasive) and surface (non-invasive) methods. They are based on the propagation of body waves and surface waves, which are associated to very small strain (<0.001%). Geophysical test data give mainly the information about the shear wave velocity (V_s) profile for a soil strata. Seismic tests are also used to determine the shear modulus and the material damping ratio by measuring the spatial attenuation of body or surface waves

$$G_{\max} = \left[\frac{\gamma}{g}\right] V_s^2 \tag{4.4}$$

$$D_o = \frac{\alpha V}{2\pi f}\ [D_0 < 10\%] \tag{4.5}$$

Where, D_o—small strain damping ratio, α—attenuation coefficient; V—velocity respectively of P, S or R waves; f—frequency.

Seismic site characterization based on shear wave velocity will provide more insight into the mechanics of site amplification and liquefaction phenomena as it is an essential and fundamental parameter for characterizing the dynamic geotechnical properties of the soil. Site characterization based on geotechnical and geophysical methods are found to be most reliable. In seismic microzonation, it is required to obtain detailed subsurface profile over the region of interest. It is difficult to carry conventional geotechnical site explorations over such a large region. In addition, carrying geotechnical site explorations over a large area is very expensive. Geophysical methods are only alternative to avoid these difficulties. These methods provide lateral variability of the near-surface materials beneath a site. Shear wave velocity (V_s) is an essential parameter for evaluating the dynamic properties of soil in the shallow subsurface. A number of geophysical methods have been proposed for near-surface characterization and measurement of shear wave velocity by using a great variety of testing configurations, processing techniques, and inversion algorithms.

The most widely used borehole methods are P-S logging, down-hole logging, and cross-hole logging [207, 279]. In down-hole logging, the travel time used by vertically propagating shear-waves from a source on the surface to a subsurface receiver along a borehole is measured. Cross-hole logging is based on subsurface measurements in which the travel time is measured by horizontally propagating shear-waves from inside a borehole to neighboring boreholes [117]. Cross-hole logging also has the advantage of identifying the properties of the soil deposit between the boreholes.

Surface Wave Tests

This method is used to measure the shear wave profiles of soil. This method depends on the dispersive characteristics of Rayleigh waves traveling through layered soil medium. A dynamic source is used to create surface waves of different frequencies and these are monitored by two receivers at known distances. Using the SASW methods large area can be covered and the soil profile can be obtained. Spectral Analysis of Surface Wave (SASW) method [95, 347] and Multichannel Analysis of Surface Wave (MASW) method [258, 388] are developed to estimate shear wave velocity profile from surface wave energy. Surface wave methods is non-invasive field tests that are executed from the ground surface without drilling any boreholes. Surface wave methods are increasingly used in civil engineering applications to evaluate soil shear modulus with depth [251]. This approach has advantages over invasive subsurface measurements [389], because it can be easily implemented along linear sections to obtain a two-dimensional shear-wave velocity profile of shallow layers [124, 195]. It can be used as a tool for imaging subsurface heterogeneity [325]. The number of surface-wave profiling applications is growing, but there are questions about experimental and theoretical limitations [251]. MASW system consists of number of geophones (usually twelve or more) against two geophones in SASW. The seismic waves are created by an impulsive source (sledge hammer). These waves are captured by the geophones. The captured Rayleigh wave is further analyzed using suitable software to generate Vs data. Both SASW and MASW involve three steps, (1) data acquisition, (2) construction of a dispersion curve, and (3) inversion of the dispersion curve to get shear wave (Vs) profile.

The term "Multichannel record" indicates a seismic data set acquired by using a recording instrument with more than one channel using geode seismograph. Both SASW and MASW effectively use the highest signal-to-noise ratio (S/N) of surface waves. The surface waves having different frequencies are generated together due to impact of sledge hammer over soil half space. However, due to the variation of engineering properties of soil mass along the depth, the surface waves of different frequency will reach geophones at different time. Knowing the distance between source and geophones, as well as geophone spacing, it is possible to generate a plot which shows the variation of velocities of the waves of different frequency. This plot is typically called a dispersion curve and its pattern depends upon the engineering properties of the strata below.

Earthquake hazards like ground motion amplification and soil liquefaction, which are responsible for major catastrophes, mainly depend on the topography and dynamic geotechnical properties of soil overlying the bedrock. The physical properties of soil such as shear wave velocity (V_s), dynamic shear modulus (G), Poisson's ratio (μ), and damping ratio (β) are referred to as dynamic geotechnical properties. The site characterization based on these properties will provide more insight into the mechanics of these hazards. Nowadays, average shear wave velocity for top 30 m (V_{s30}) has been widely accepted for seismic site characterization. It provides an unambiguous site characterization and permits quantitative site amplification [39]. Most of the codes like [86], NEHRP [51], International Building Code [130], etc. specify the site characterization based on the average shear wave velocity values in the top 30 m (V_{s30}) evaluated using Eq. (4.6).

$$V_{s30} = \frac{30}{\sum_{i=1}^{n} \frac{d_i}{V_i}} \tag{4.6}$$

where d_i is the thickness of individual layers and V_i is the shear wave velocity through each layer and 'n' is the number of layers. Many literature show the use of the average shear wave velocity values in the top 30 m (V_{s30}) has become an important parameter for site classification and it is being incorporated in various codes like [86], NEHRP [51], International Building Code [130], etc. The phenomenon of site amplification is strongly influenced by the near surface shear wave velocity which can be better represented by the average shear wave velocity in the top 30 m. Table 4.5 describes the NEHRP site classification based on V_{s30}.

A site classification scheme based on V_{s30} values was proposed by Borcherdt [39] and a similar scheme was adopted by the National Earthquake Hazard Reduction Program (NEHRP) also. The NEHRP [51] site classification scheme is given in Table 4.5. Eurocode-8 [86] has also classified the site based on V_{s30}, standard penetration test (SPT), and cone penetration test (CPT) values. The classification given by Eurocode-8 is given in Table 4.6.

In many locations, the rock depth will be shallow (less than 30 m) and hence the evaluation of V_{s30} value will not be possible. In those cases, extrapolation of

Table 4.5 Site classification as per NEHRP scheme [51]

NEHRP site class	Description	V_{s30}
A	Hard rock	>1500 m/s
B	Firm and hard rock	760–1500 m/s
C	Dense soil, soft rock	360–760 m/s
D	Stiff soil	180–360 m/s
E	Soft clays, special study soils, e.g. liquefiable soil	<180 m/s

Table 4.6 Site characterization as per [86]

Ground type	Description of stratigraphic profile	V_{s30} (m/s)	SPT	Cu (kPa)
A	Rock or other rock-like geological formation, including utmost 5 m of weaker material at the surface	>800		
B	Deposits of very dense sand, gravel, or very stiff clay, at least several tens of meters in thickness, characterized by a gradual increase of mechanical properties with depth	360–800	>50	>250
C	Deep deposits of dense or medium dense sand, gravel or stiff clay with thickness from several tens to many hundreds of meters	180–360	15–50	70–250
D	Deposits of loose-to-medium cohesion less soil (with or without some soft cohesive layers), or of predominantly soft-to-firm cohesive soil	<180	<15	<70
E	A soil profile consisting of a surface alluvium layer with V_{s30} values of type C or D and thickness varying between about 5 and 20 m, underlain by stiffer material with $V_{s30} > 800$ m/s			
S1	Deposits consisting, or containing a layer at least 10 m thick, of soft C or D and thickness varying between clays/silts with a high plasticity index (PI > 40) and high water content	<10 (indicative)		10–20
S2	Deposits of liquefiable soils, of sensitive clays, or any other soil profile not included in types A-E or S1			

available V_s values has to be done to evaluate the V_{s30} values. The method proposed by Boore [36] can be used for this purpose. He has suggested different models to extrapolate the shear wave velocities, for depths less than 30 m, to get the V_{s30} value. The first method is extrapolation based on constant velocity. In this model it is assumed that the shear wave velocity remains constant from the deepest velocity measurement to the 30 m.

$$V_{s30} = \frac{30}{tt(d) + \frac{(30-d)}{V_{eff}}} \tag{4.7}$$

where $tt(d)$ is the travel time to depth d and $V_{eff} = V_{s(d)}$, $V_{s(d)}$ is the timed average velocity to a depth of d.

Even though this method is simple, it is found to underestimate the V_{s30} values, since in most of the soils, the shear wave velocity is found to increase with depth. Another relation proposed by Boore [36] was based on a power law relation, the V_{s30} value can be estimated as given in equation

$$\log V_{s30} = a + b \log \overline{V}_{s(d)} \tag{4.8}$$

where $\overline{V}_{s(d)}$ is the velocity at a depth of d m ($10 < d < 30$). The values of the regression coefficients a and b can be obtained from [36]. The extrapolation of V_s values can also be done based on the velocity statistics [36].

$$P\left(\xi > \frac{V_{eff}}{V_s(d)}\right) = a\left(\frac{V_{eff}}{V_s(d)}\right)^b \tag{4.9}$$

where $P\left(\xi > \frac{V_{eff}}{V_s(d)}\right)$ is the probability of exceedance of $\left(\frac{V_{eff}}{V_s(d)}\right)$. More details of this analysis can be had from [36].

A modified site classification system based on geotechnical data was proposed by Rodriguez-Marek et al. [290]. In this, the stiffness of soil was also taken into account for the site classification as shown in Table 4.7. The main advantage of this system is that it correlates the V_{s30} values with the geotechnical and surface geological features.

4.2.4 Based on the Topographic Slope

For small sites, V_{s30} parameter for site characterization can be evaluated using geotechnical and geophysical field tests. However for characterizing larger area,

Table 4.7 Classification based on geotechnical features [290]

Site	Description	Comments
A	Hard rock	Crystalline bedrock; $V_{s30} \geq 1500$ m/s
B	Competent bed rock	$V_{s30} > 600$ m/s or <6 m of soil Most unweathered California rock cases
C1	Weathered rock	V_{s30} 300 m/s increasing to >600 m/s, weathering zone >6 m and <30 m
C2	Shallow stiff soil	Soil depth >6 m and <30 m
C3	Intermediate depth stiff soil	Soil depth >30 m and <60 m
D1	Deep stiff Holocene soil	Soil depth >60 m and <200 m
D2	Deep stiff Pleistocene soil	Soil depth >60 m and <200 m
D3	Very deep stiff soil	Soil depth >200 m
E1	Medium thickness soft clay	Thickness of soft clay layer 3–12 m
E2	Deep soft clay	Thickness of soft clay layer >12 m
F	Potentially liquefiable sand	Holocene loose sand with high water table, $Z_w \leq 6$ m

estimation of V_{s30} based on these tests is not economically as well as physically viable. Generally in such cases, site characterization based on V_{s30} is carried out using available geological maps.

However geology-based maps are typically mapped with a goal other than the estimation of seismic site amplification, and hence the use of these maps for first order site characterization always associates with certain drawbacks [379]. In site condition map derived from geological map, a single value of V_{s30} is assigned to an individual geological unit and it does not capture the variability of V_{s30} within that unit. Studies have proved that the topographic variations are an effective index of near-surface geomorphology and lithology, with steep mountains representing rocky terrain and flat basins indicating soil, and intermediate slopes representing a transition between rock and soil [379]. The topographic-slope more accurately maps the variation of V_{s30} across a geological unit when compared with a geological map, by characterizing the presumed change in particle size with topographic gradient. Matsuoka et al. [211] have confirmed that good correlations exist between V_{s30} and slope and geomorphic indicators in Japan. A study was done by Chiou and Youngs [67] in Taiwan and they found that a good correlation exists between elevation and V_{s30}.

Wald and Allen [379] have described a technique to derive first-order site-condition maps directly from topographic data. They have used global 30 arc sec topographic data and V_{s30} measurements collected from several studies in the United States, Taiwan, Italy, and Australia. They have correlated V_{s30} values with the topographic slope to develop two sets of parameters for obtaining V_{s30}: one for active tectonic regions where the topography is steep, and one for stable shield regions where the topography is gentler. By taking the gradient of the topography and choosing ranges of slope that maximize the correlation with shallow shear-velocity observations; [379] have recovered, to first order, many of the spatially varying features of the site-condition maps developed in California (see Fig. 4.3). They found that the maps derived from the slope of the topography were correlating well with other independently derived, regional site-condition maps. Similar observations were also found in the case of other regions such as Los Angeles, Taiwan, Salt Lake City, Utah and Memphis, Tennessee [379]. Hence from all the available literature it is inferred that a reliable first order site condition map based on V_{s30} can be derived from topographic slope maps and hence has been adopted in the present study.

James and Sitharam [151] has carried out a comparison study between V_{s30} maps at the meso-level derived from the topographic slope data and V_{s30} maps prepared by various researchers using geotechnical and geophysical testing. Figure 4.4 presents the comparison between V_{s30} map of Bangalore prepared by Sitharam and Anbazahgan [329] and V_{s30} map derived from the topographic slope data. From Fig. 4.4a, b it is clear that the western part of Bangalore is in the site class C and B with V_{s30} in the range of 400–900 m/s.

Similarly the eastern part of Bangalore mainly constitutes of site class D and C with V_{s30} varying from 200 to 600 m/s. Some differences in the distribution of V_{s30} values can be observed in the south-central part of Bangalore. As per [329], this part

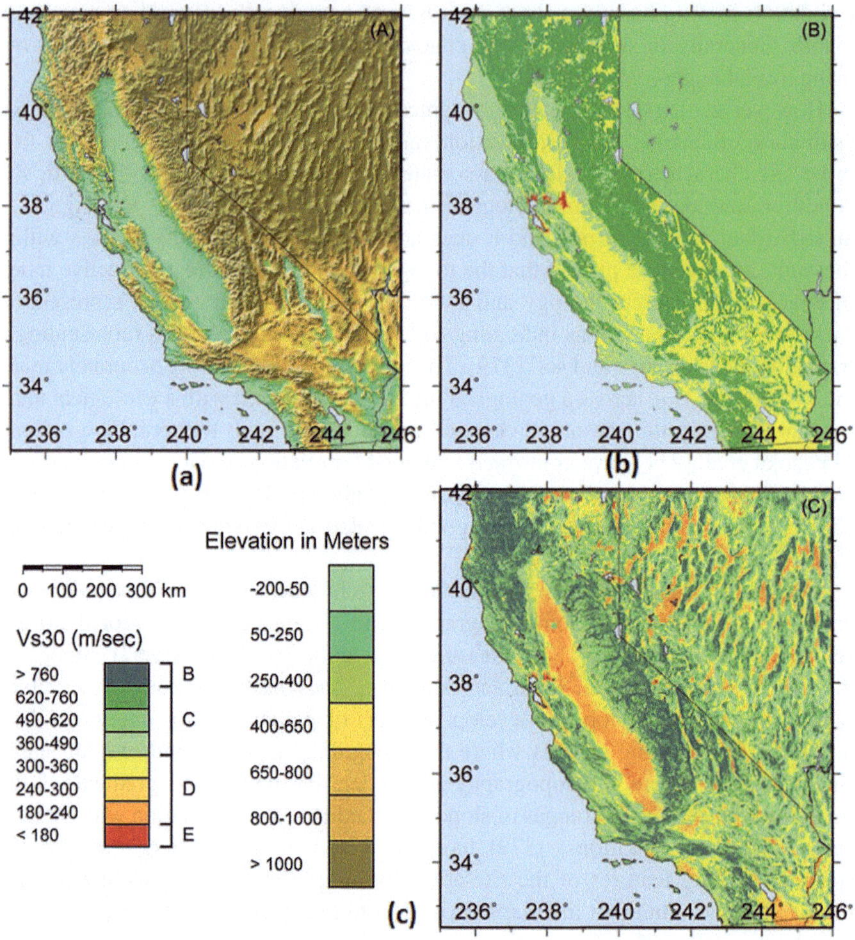

Fig. 4.3 (**a**) The topographic relief for the state of California, (**b**) Site class map for California [384], (**c**) Site class map for California derived from topographic slope map [379]

falls in the site class C and B with V_{s30} values ranging from 400 to 900 m/s, while in the present study this region comes under site class D to C (V_{s30} ranging from 200 to 600 m/s).

James and Sitharam [151] has also carried out a similar comparative study for the city of Chennai (Madras) also. The V_{s30} map derived from topographic slope map (Fig. 4.5b at a grid size of 1 km × 1 km) was compared with V_{s30} prepared by Uma Maheswari et al. [367] using MASW test data (Fig. 4.5a). Both the figures show that the major area of the city falls in site class D, with V_{s30} ranging between 180 and 360 m/s. The deviation between two maps can be observed in the southern parts of the city, which is classified as site class C and B as per [367], while the present study shows the area fall in site class D and C.

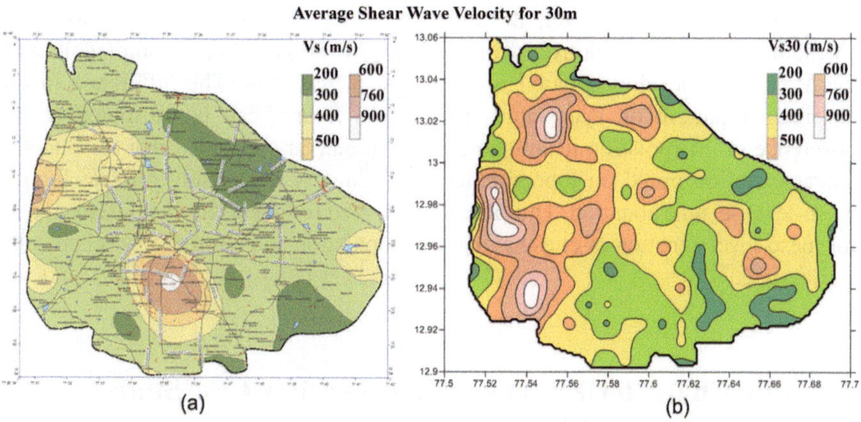

Fig. 4.4 Comparison of V_{s30} maps obtained using two different methodologies for Bangalore (a) developed by Sitharam and Anbazahgan [329] (b) derived from the topographic slope data (after [151])

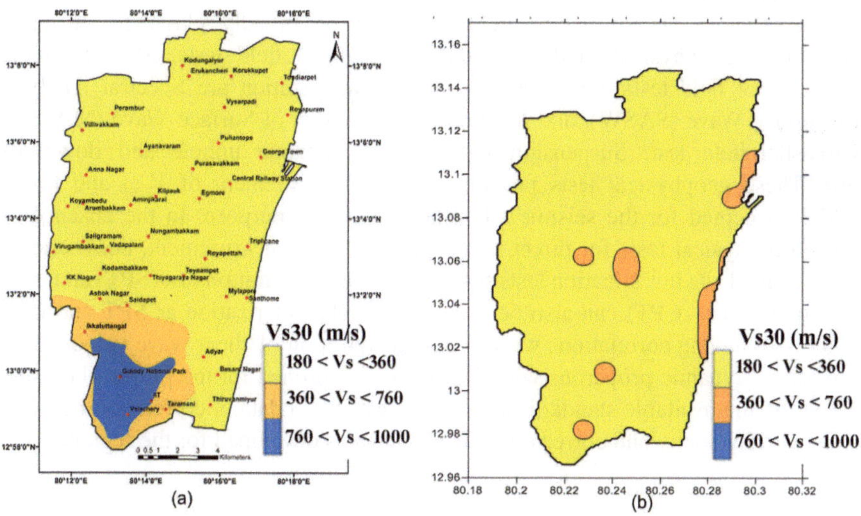

Fig. 4.5 Comparison of V_{s30} maps obtained using two different methodologies for Chennai (a) developed by Uma Maheswari et al. [367] (b) derived from the topographic slope data (after [151])

However in a broad level, it can be observed that for both Bangalore and Chennai, the V_{s30} maps derived from topographic slope data are comparable with the V_{s30} maps developed based on geophysical test data. The difference in the distribution pattern of V_{s30} values in the two figures can be attributed to the difference in interpolation techniques. Hence it can be concluded that seismic site characterization based on topographic slope maps is moderately reliable for the meso-level seismic zonation. Thus it can be adopted for macro-level studies and also for meso-level zonation of cites having low population and lying in low seismically active zone.

4.3 Site Characterization at Micro-Level: A Case Study of Kalpakkam NPP Site

James et al. [153] has carried the seismic site characterization for the Kalpakkam NPP site using geotechnical/geophysical tests. For small region such as the Kalpakkam NPP site, various in-situ field tests are available for seismic site characterization. These in-situ field tests can be broadly classified into two categories, geophysical field tests and geotechnical field tests. Most common geophysical field tests conducted for site characterization are Spectral Analysis of Surface Wave (SASW), and Multichannel Analysis of Surface Wave (MASW), Cross-borehole test, Suspension logging test, Seismic uphole and downhole test. These geophysical tests provide a direct measurement of V_{s30} and hence widely accepted for the seismic site characterization purpose. In the absence of these geophysical tests for direct shear wave velocity measurement, conventional geotechnical site investigation tests such as standard penetration test (SPT) and cone penetration test (CPT) can also be used for site characterization as SPT and CPT values are having correlations with V_{s30}. The variation of shear wave velocity and low strain dynamic properties with depth was determined for the power plant site. Based on the available standard penetration test (SPT) data, a correlation between shear wave velocity and corrected SPT-N value was developed for the region.

4.3.1 MASW Testing for Site Characterization

Shear wave and P-wave velocity (V_s, V_p) are very well related to the dynamic properties of soil at lower strains. The most widely used technique for the measurement of shear wave and P-wave velocity is by MASW (Multichannel Analysis of Surface Waves) technique. MASW is a non-destructive, indirect geophysical method, in which the variation of shear-wave velocity with depth is obtained by analyzing the surface wave propagation pattern, which is recorded using multiple channels. The MASW method was first introduced by Park et al. [258]. The surveying

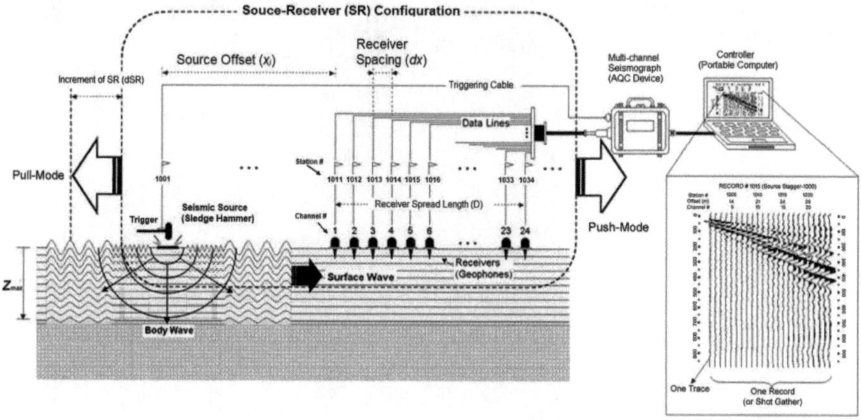

Fig. 4.6 Schematic showing typical active MASW survey (taken from http://www.masw.com/DataAcquisition.html)

procedure consists generation of surface waves using an active seismic source (e.g., a sledge hammer) and capturing generated surface wave using a linear receiver array as shown in Fig. 4.6. The MASW is used in geotechnical engineering for soil stratum identification based on shear wave velocity variations along the depth. It involves three steps: data acquisition, construction of a dispersion curve (showing the variation of phase velocity among various frequencies), and inversion of the dispersion curve to get shear wave (V_s) profile from the calculated dispersion curve. The typical MASW test setup used for the present study consists of 24-channel geophones of 4.5 Hz capacity as shown in Fig. 4.7. A propelled energy generator (PEG) (Fig. 4.8) was used as active source for generating surface waves instead of a conventional sledge hammer. The PEG was found to be more efficient than the conventional sledge hammer for generating surface wave. The PEG consists of a 40 kg hammer dropped from a height of 36–40 cm. The PEG is designed to easily mount on trucks or trolley and can be conveniently moved to various locations.

Spatial distribution of MASW test locations in the site is shown in Fig. 4.9 which were selected based on its proximity to certain borehole locations. A typical record obtained using 24 channel geophones is presented in Fig. 4.10. The acquired signal was analyzed using Surfseis software [259] and a dispersion image showing the distribution of signal intensity and phase velocity among various frequency components of surface waves was generated. The conventional spectral analysis of surface waves (SASW) method measures the phase difference (ϕ_{diff}) of surface waves for a frequency (f) between the two receivers. The phase velocity (C_f) of each frequency was then calculated from the relationship: $C_f = \frac{2\times\pi\times f}{\phi_{diff}}$. This procedure is repeated for different frequencies to construct a dispersion curve. However, in the MASW analysis, instead of calculating the individual phase velocity for each frequency, an image space (Fig. 4.11) was constructed showing

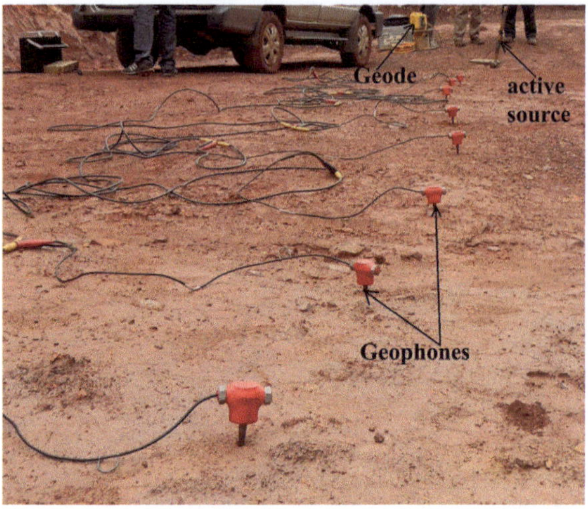

Fig. 4.7 Typical MASW setup in the field

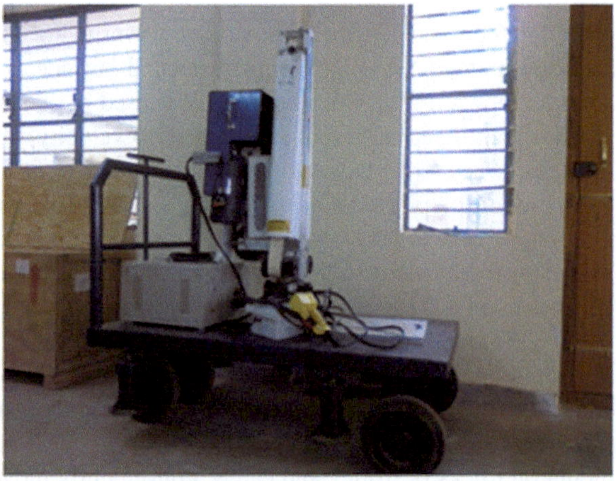

Fig. 4.8 Propelled energy generator assembly in the MASW test (after [150])

the dispersion trends which were identified from the pattern of energy accumulated in this space. The phase-shift method proposed by Park et al. [257] was used for the dispersion imaging scheme and is incorporated in the Surfseis software. Then, the necessary dispersion curves are extracted by following the image trends. A dispersion curve was fitted to this dispersion image (Fig. 4.11), through regions where the intensity of signal was found to be high (high signal to noise ratio). Finally the shear wave (V_S) variation below the surveyed area, which is most responsible for the analyzed dispersion pattern of surface waves, was deduced using an inversion technique. Inversion methodology proposed by Xia et al. [388] is incorporated in the Surfseis software. An initial earth model is assumed at the start of the inversion process which consists of information like shear and P wave velocity, density,

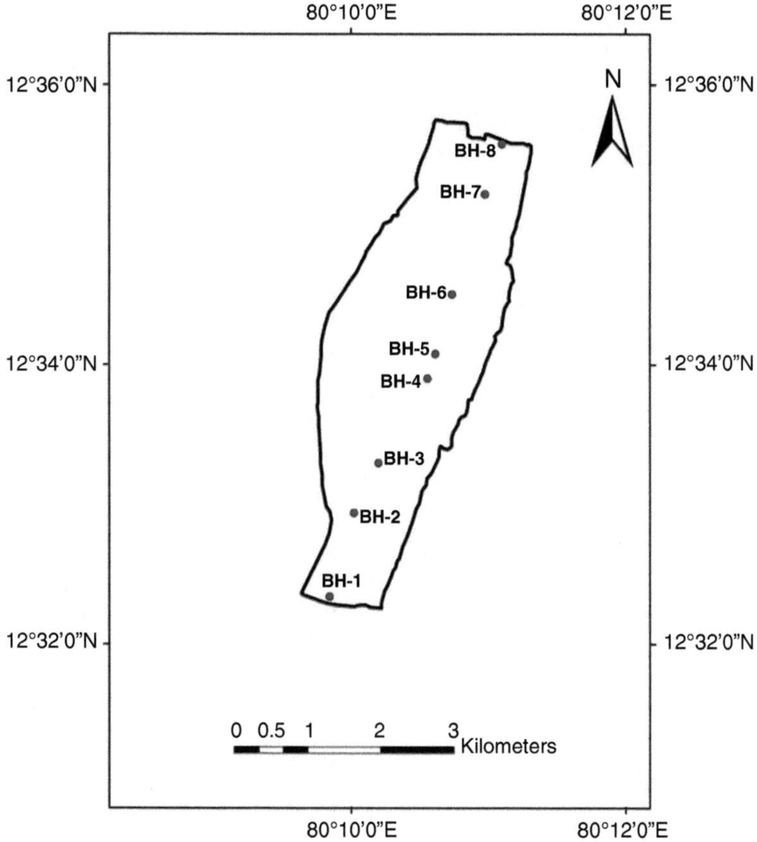

Fig. 4.9 Locations of control point where MASW testing was carried out (after [150])

Poisson's ratio, and layer thickness. Of these parameters the shear wave velocity is most significant and influential parameter which controls the changes in Rayleigh-wave phase velocities for a layered earth model [388]. Hence keeping the other parameters such as P wave velocity, density, Poisson's ratio, and layer thickness as constant, the shear wave velocity is updated in the initial earth model after each iteration till the theoretical dispersion curve matches with the actual dispersion curve obtained from the field. With known (or assumed) value of Poisson's ratio, the variation of P-wave velocity with depth (Fig. 4.12) is also being obtained.

As per the studies of [39, 274, 345], the average shear wave velocity for 30 m is decisive for site characterization, was evaluated as per Eq. (4.6). It is evident from Fig. 4.13 that the site is mainly categorized into two major site classes, site class C (dense soil or very soft rock) and site class D (stiff soil) on the basis of average shear wave velocity for top 30 m as per NEHRP [51].

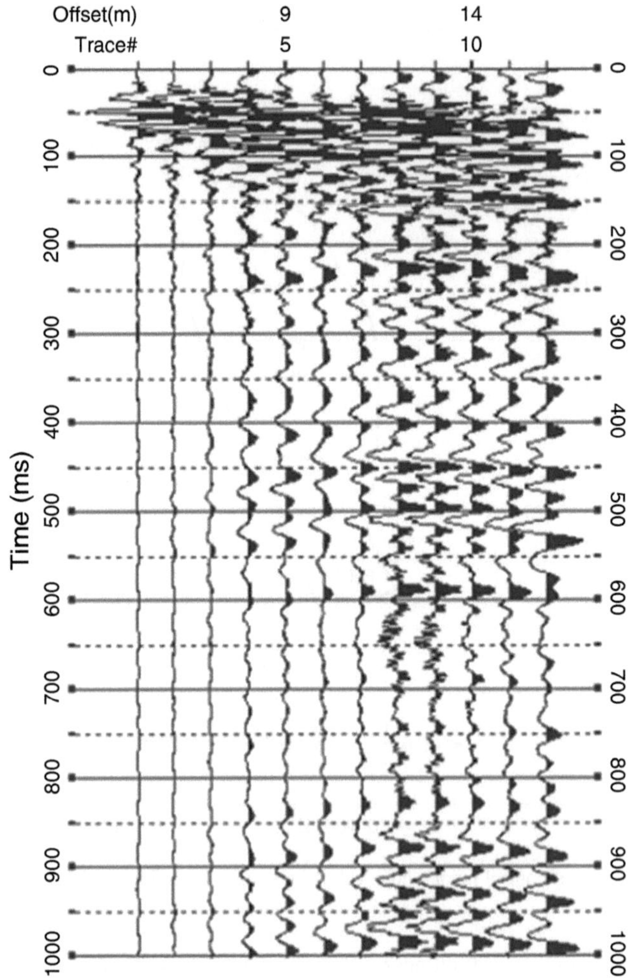

Fig. 4.10 Typical MASW recorded seismogram (after http://www.masw.com/DataAcquisition.html)

4.3.2 Evaluation of Low Strain Dynamic Properties

Low strain dynamic properties of soil includes shear modulus, damping ratio, and Poisson's ratio. Shear modulus (G) is an important soil parameter required to predict site response and liquefaction potential. Shear modulus (G) of any soil material subjected to cyclic loading (provided constant density and confining pressure) depends upon the magnitude of shear strain induced in the specimen [140, 181]. At strain level well below 10^{-5}, soil exhibits elastic behavior and the stiffness (shear modulus) of the soil at this strain range is independent of loading rate and number of load repetitions [140]. Thus the shear modulus will have its maximum

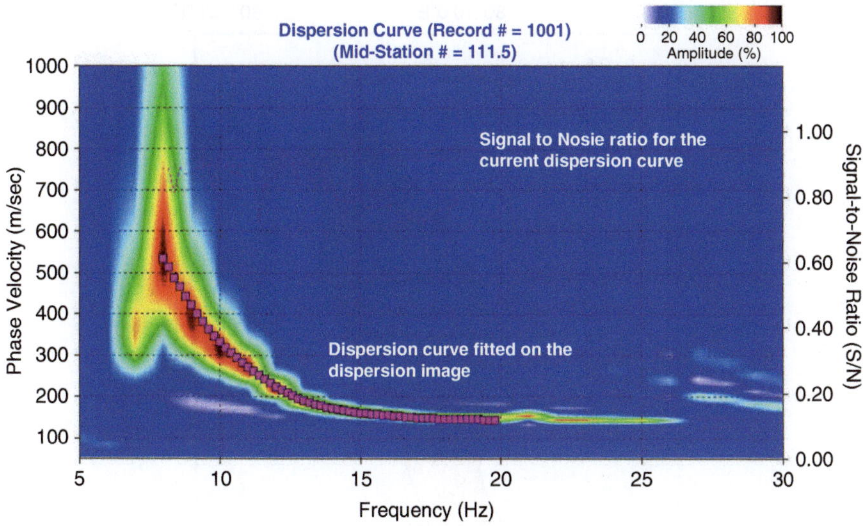

Fig. 4.11 Typical dispersion image obtained after analyzing MASW data (after [153])

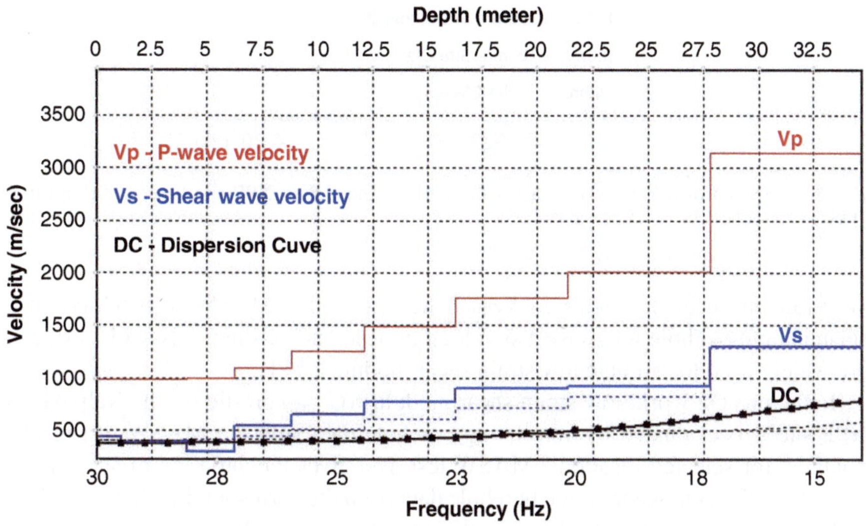

Fig. 4.12 Typical Shear wave and P-Wave velocity profile of a location in the study area (after [153])

value at a strain level less than 10^{-5} (for samples having a constant density and confining pressure) which ideally represents the in situ conditions. The degradation of shear modulus occurs with the increase in the strain level as represented by a modulus reduction curve which is an essential input for equivalent linear ground response analysis. As the phenomenon of vibrations and wave propagation induces

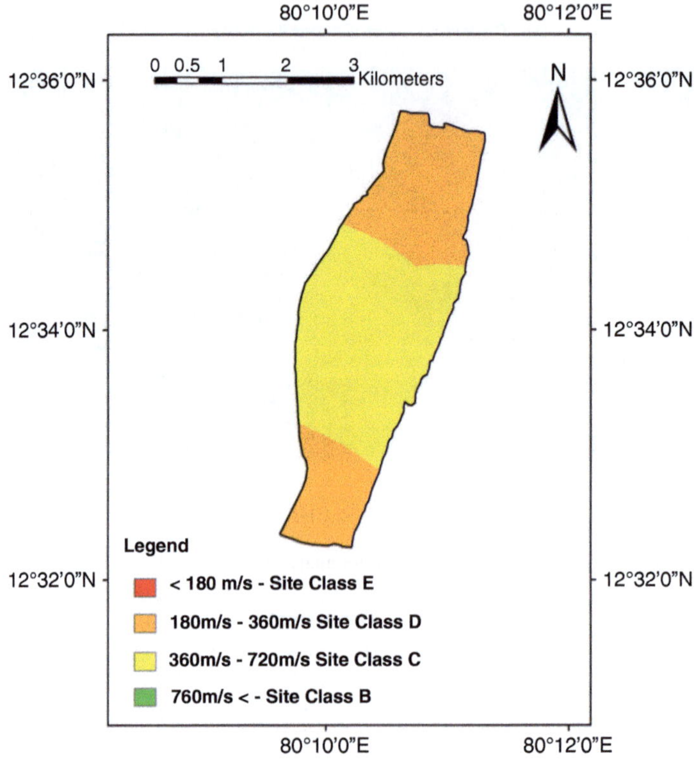

Fig. 4.13 Spatial distribution site classes throughout Kalpakkam NPP site based on V_{s30} (after [153])

very low strain in the soil, geophysical field tests like MASW test, SASW test, seismic up/downhole test, suspension logging test, and seismic cross-hole test, etc. are widely used to estimate low strain shear modulus [181].

In the present work, low strain shear modulus ($G_{\max}$) profile for the Kalpakkam NPP site was evaluated by correlating borehole data with the MASW test results obtained for selected locations. MASW test data gives the variation of shear wave velocity with depth whereas the borehole data from the corresponding location gives soil type and in-situ density values with depth. Low strain shear modulus ($G_{\max}$) is obtained from Eq. (4.10).

$$G_{\max} = \left[\frac{\gamma}{g}\right] V_s^2 \tag{4.10}$$

where γ is the in-situ density (in kN/m^3), g—acceleration due to gravity (in m/s^2), and V_s is the shear wave velocity (in m/s). Figures 4.14 and 4.15 show the shear wave velocity and P-wave velocity variation with depth for eight major locations in the power plant site.

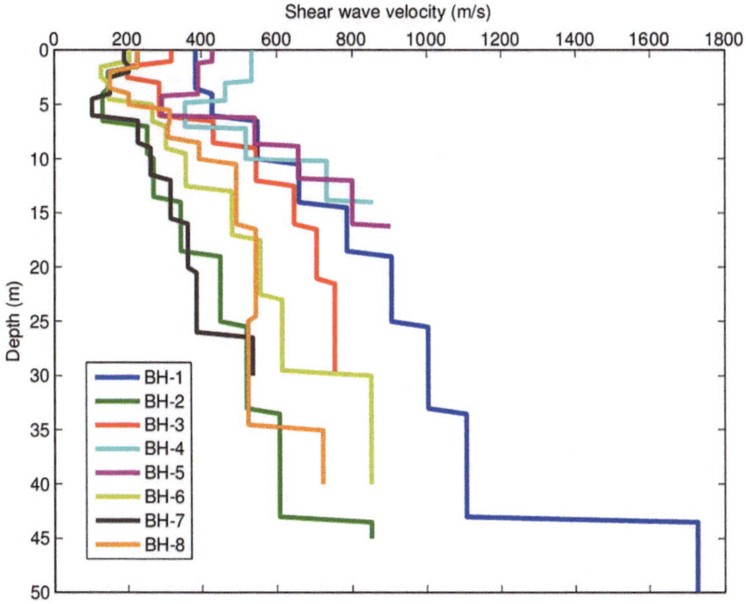

Fig. 4.14 Shear wave velocity profile for selected locations in the study area (after [150])

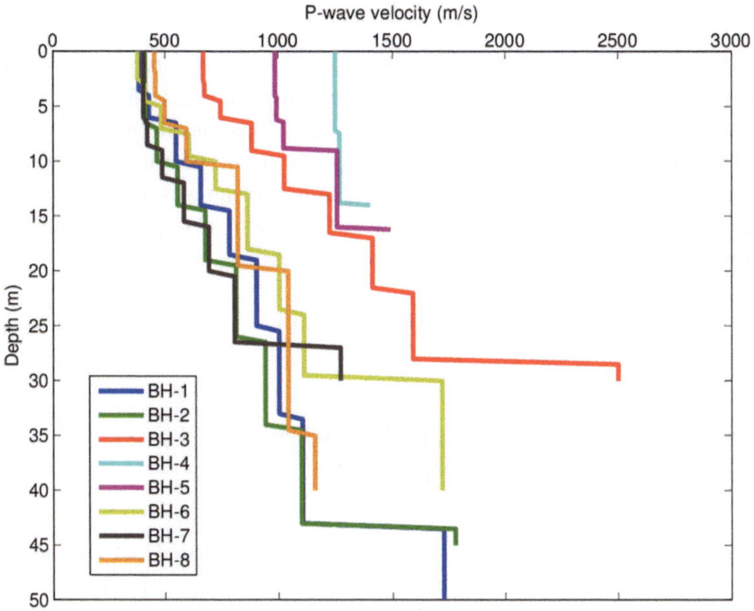

Fig. 4.15 P-wave velocity profile for selected locations in the study area (after [150])

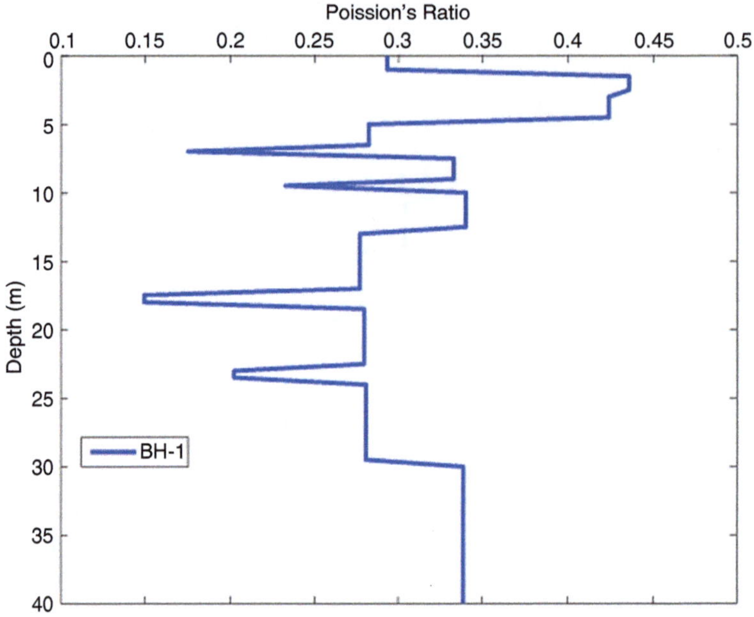

Fig. 4.16 Typical plot showing the variation in Poisson's ratio with depth (at BH-1) (after [150])

The Poisson's ratio is obtained from the MASW test and its typical variation with depth at a given location in the study area is presented in Fig. 4.16. Low strain shear modulus ($G_{\max}$) profile with depth for eight major locations in the study area were evaluated and presented in Fig. 4.17. Knowing low strain shear modulus ($G_{\max}$) and Poisson's ratio (ν), Young's modulus ($E_{\max}$) at low strain was also evaluated using the standard elasticity relation given in Eq. (4.11). Figure 4.18 presents the variation of Young's modulus with depth for eight major locations in the NPP site.

$$E = 3G \times (1 + \nu) \tag{4.11}$$

4.4 Site Characterization at Macro-Level: A Case of Karnataka State

This section presents the seismic site characterization carried out for the state of Karnataka, India. It is to be noted that for characterizing large area, estimation of V_{s30} based on geotechnical/geophysical tests is not economically as well as physically viable. Hence, James et al. [151] have carried out site characterization for the state of Karnataka using the topographic slope map generated from digital elevation model (DEM) data.

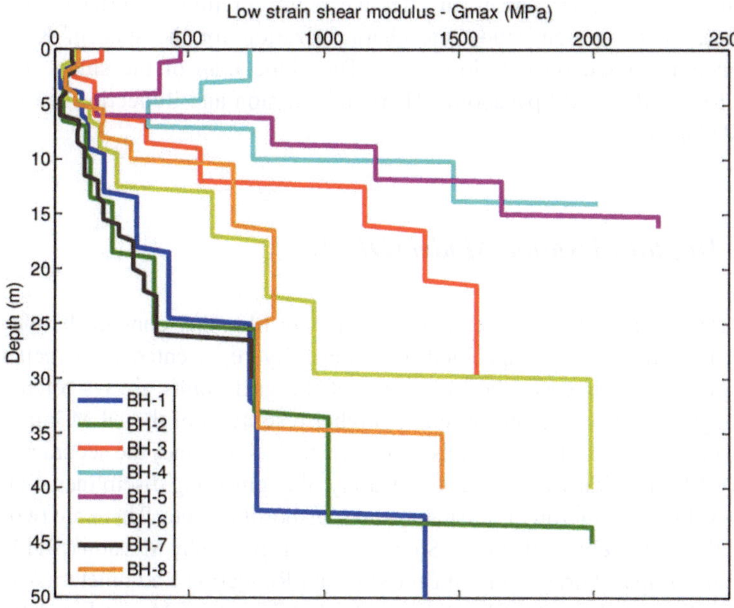

Fig. 4.17 Low strain shear modulus (G_{max}) profile for selected locations in the study area (after [150])

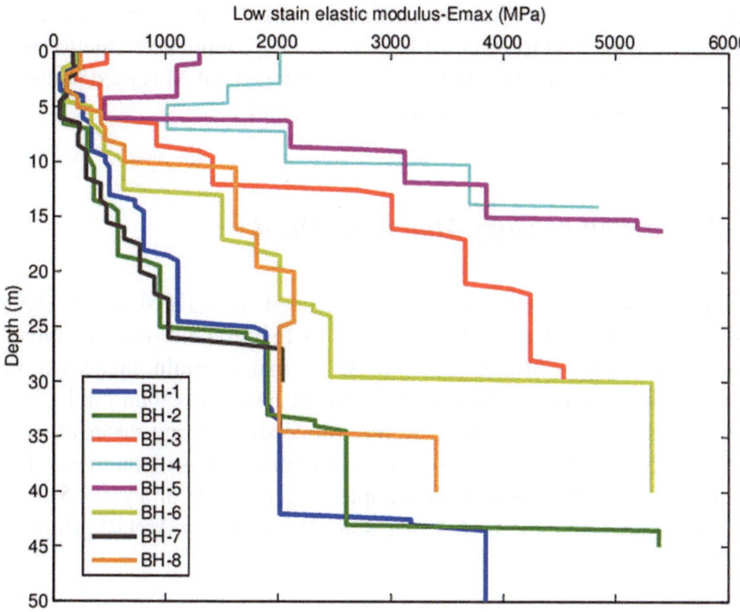

Fig. 4.18 Low strain elastic modulus (E_{max}) profile for selected locations in the study area (after [150])

As the terrain topography is an effective representation of surface geological conditions, in the present study site characterization for the state of Karnataka has been done based on the slope map. The slope map of the study area was derived from Advanced Spaceborne Thermal Emission and Reflection Radiometer (ASTER) DEM.

4.4.1 Digital Elevation Model (DEM)

Digital elevation models are generally raster data files that contain the elevation data of the terrain over a specified area generally represented as a rectangular grid of pixels. The intervals between each of the grid points always reference to some geographical coordinate system. A higher resolution of the DEM implies the distance between two adjacent pixels is small, which determines the accuracy of that DEM. DEMs are widely used in a large variety of engineering disciplines such as in hydrology, landslide hazard assessment, and transportation, etc. There are two kinds of DEM that are freely available (1) Shuttle Radar Topography Mission (SRTM) and (2) Advanced Spaceborne Thermal Emission and Reflection Radiometer (ASTER). The ASTER Global Digital Elevation Model (ASTER GDEM) (in Fig. 4.19) was developed and released by the Ministry of Economy, Trade, and Industry (METI) of Japan, in collaboration with the National Aeronautics and Space Administration (NASA) United States. The ASTER GDEM covers 99% of the total land area, spanning from 83°N to 83°S with a resolution of 1 arc–(30 m). The estimated vertical accuracy for ASTER GDEM is 20 m at 95% confidence. Due to such high resolution for a freely available DEM data, the ASTER GDEM is used in the present study to generate a slope map for the state of Karnataka.

4.4.2 Extraction of Slope Map from DEM

Slope map for the two study areas were developed from DEM using ArcGIS 10 software. ArcGIS 10 is the most advanced GIS (geographic information system) software developed by ESRI, United States. All major terrain analysis in present study was carried out using ArcGIS. The ASTER DEM was found to be in WGS-1984 geographic coordinate system. Before proceeding to any raster analysis, the DEM was projected to the UTM co-ordinate system which is recommended for any terrain analysis. Then the DEM was resampled to a grid size of 5 km $\times$ 5 km for the state of Karnataka. Using the spatial analyst tool provided by ArcGIS, slope at each

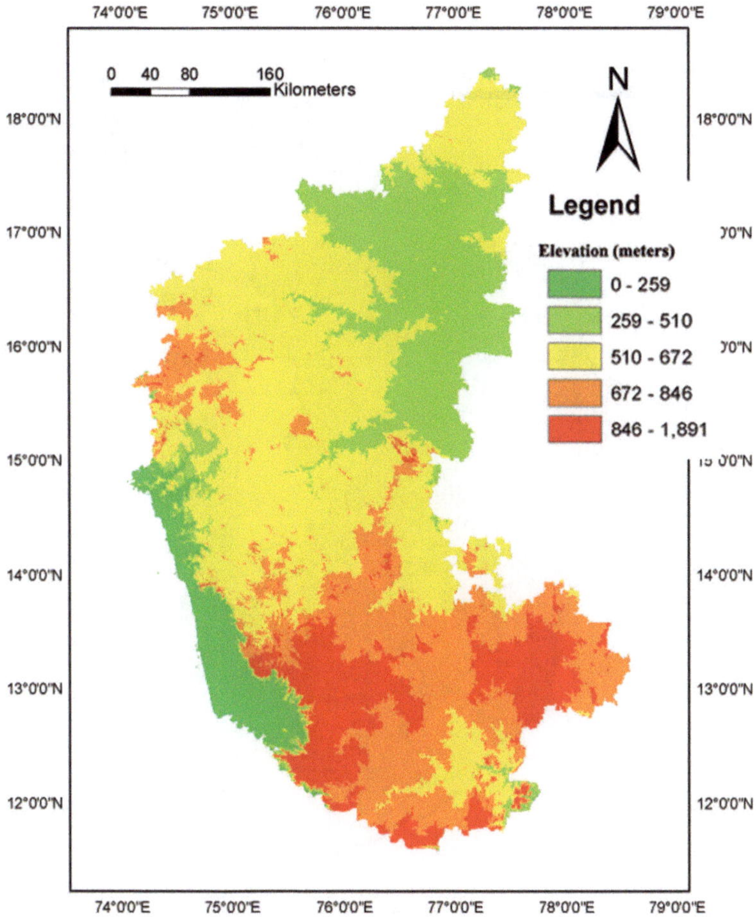

Fig. 4.19 Digital elevation model (DEM) of the Karnataka state (after [151])

grid point was evaluated. Slope map of the Karnataka state is presented in Fig. 4.20. Slope map thus obtained was converted to vector data (points) and then exported into a table form. For the state of Karnataka, there were 8691 grid points for which geographic co-ordinates and slope values were computed.

4.4.3 Site Characterization Based on Terrain Slope

Once slope maps are obtained, site characterization for the study area was carried out using methodology suggested by Wald and Allen [379] as mentioned in Sect. 4.2.4 of Chap. 4. Based on the correlation studies conducted for active tectonic and stable continental regions, Wald and Allen [379] has proposed slope ranges

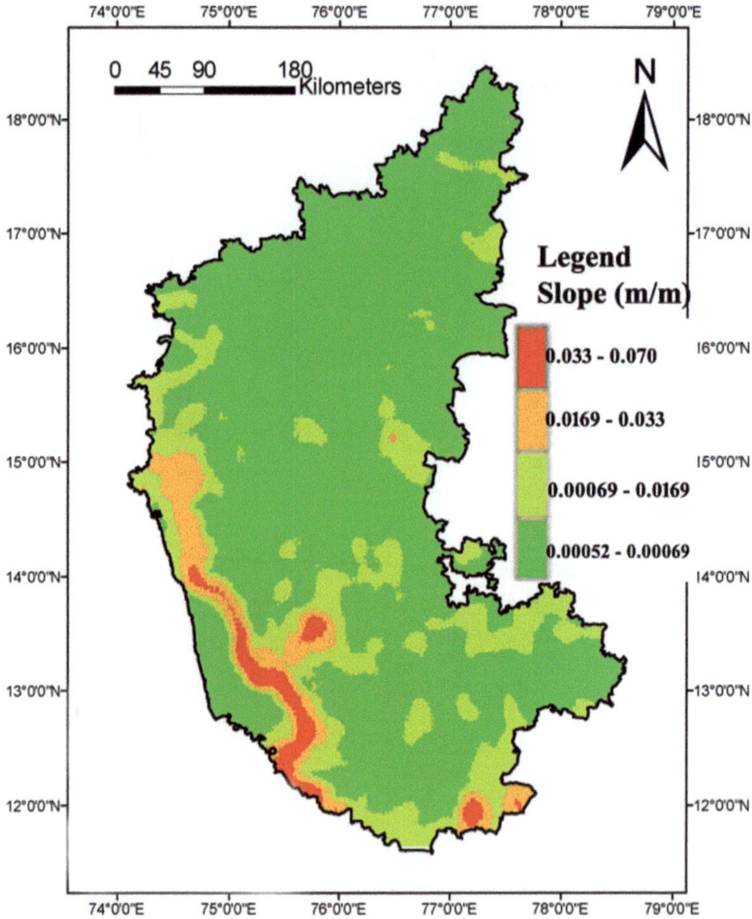

Fig. 4.20 Slope map of the Karnataka state (after [151])

corresponding to each NEHRP site class as given in Table 4.8. Based on the slope range given in Table 4.8, site class for each grid point was determined. Maps showing the spatial variation in site classes throughout Karnataka was developed and presented in Fig. 4.21. A similarly methodology has been adopted by Sitharam et al. [335] to develop the site class distribution map of entire India.

Table 4.8 Slope ranges for NEHRP site classes (after [379])

Site class (as per NEHRP)	V_{s30} range ($m\ s^{-1}$)	Slope range (m/m)	
		Active tectonic	Stable continent
E	<180	<1.0E−4	<2.0E−5
D	180–240	1.0E−4–2.2E−3	2.0E−5–2.0E−3
	240–300	2.2E−3−6.3E−3	2.0E–3−4.0E−3
	300–360	6.3E−3−0.018	4.0E−3–7.2E−3
C	360–490	0.018–0.05	7.2E−3–0.013
	490–620	0.050–0.10	0.013–0.018
	620–760	0.10–0.138	0.018–0.025
B	>760	>0.138	>0.025

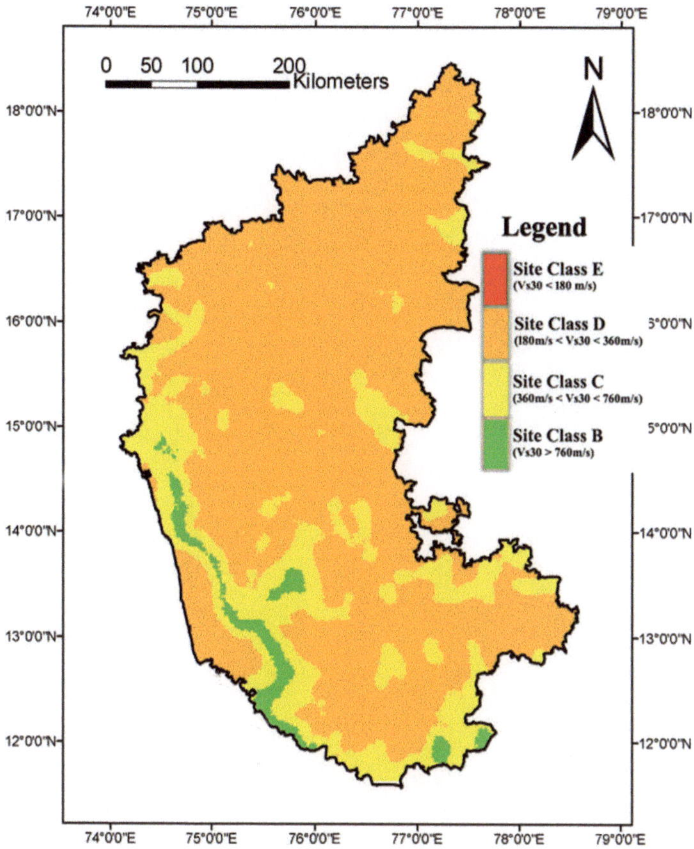

Fig. 4.21 Spatial distribution of V_{s30} values throughout the Karnataka state (after [151])

Chapter 5
Local Site Effects for Seismic Zonation

5.1 Introduction

It has been evident from the past earthquake events all over the world that the amplification of ground motion is highly dependent on the local geological, topography, and geotechnical conditions. Many researchers have worked on the estimation of local site effects [265, 266, 317, 338, 345, 357, 383]. It is observed that large concentration of damage in specific areas during an earthquake is due to site dependent factors related to surface geological conditions and local soil altering seismic motion. After the 1923 Kanto earthquake, it is very clear that major seismic damage was controlled by local geology. It is well known that each soil type responds differently, when subjected to the ground motions, imposed due to earthquake loading. Damages caused by 1985 Michoacan earthquake (M_S = 8.1), 1989 Loma Prieta earthquake (M_S = 7.1) and 2001 Bhuj earthquake (M_W = 7.7) have all reaffirmed that apart from the source characteristics, the local soil conditions also greatly influence the damage potential of an earthquake [181, 231]. The predominant influence of local soil geology/conditions on the earthquake ground motion characteristics is termed as local site effects, which have a direct impact on the response of structures during each of these earthquake events.

Hence, considering the effect of local site in the seismic design of structural and geotechnical system will improve the performance of buildings against future earthquakes. Amplification of seismic waves is one of the major local site effects, causing extensive damage to structures due to aggravated ground shaking. The amplification of seismic waves occurs when it enters into looser medium (soil) from a dense medium (rock) due to the phenomenon of impedance contrast. This chapter explains the assessment of site amplification both at the micro and macro-levels.

The response of the soil varies from type to type, when they are subjected to the earthquake excitations. When the seismic wave propagates through different medium, it tries to conserve the energy flux whenever it enters into the new medium.

T. G. Sitharam et al., *Comprehensive Seismic Zonation Schemes for Regions at Different Scales*, https://doi.org/10.1007/978-3-319-89659-5_5

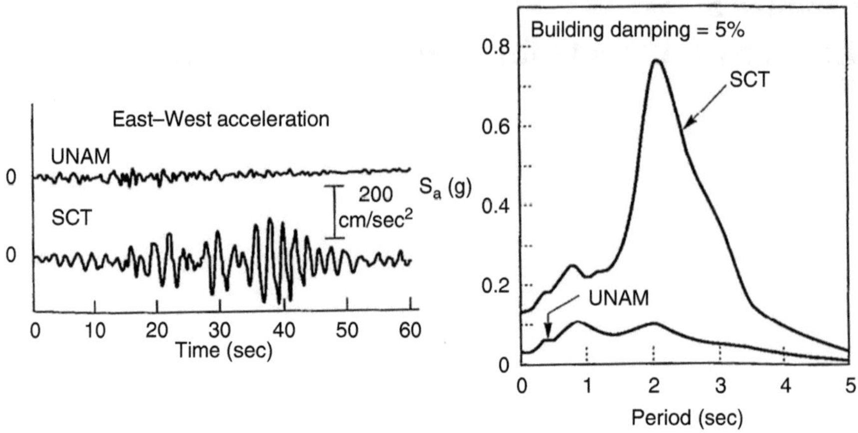

Fig. 5.1 Site amplification observed during Mexico earthquake—SCT-soft soil site and UNAM-rock site (after [348] taken from [181])

It is well established that the energy flux developed during the transmission of an earthquake wave in any medium depends upon the particle velocity and impedance (product of its density and wave velocity) of the medium. So when the wave enters a loose medium having low impedance, the particle movement gets exaggerated, so as to conserve the energy thus resulting in the amplification of seismic waves. Hence the younger loose soils often amplify the ground motion more when compared to the older compact soils or hard rock. During the 1819 Kutch earthquake, [202] observed the buildings with foundation resting on hard rock suffers less damage when compared to building founded on soil. Similar type of observations was made by Mallet [206] during the Neapolitan earthquake (of 1857), Wood [387] and Reid [281] during the San Francisco earthquake (of 1906). A well-documented case study of site effects on earthquake motion was observed during the 1985 Michoacan (Mexico) earthquake and the 1989 Loma Prieta (California) earthquake. Comparing strong motion records of 1985 Michoacan (Mexico) earthquake, at rock and soft soil site (as in Fig. 5.1) shows the amplification due to soft overlying soil [348].

5.2 Effect of Local Site Conditions on Ground Motion

Many historical evidences of the influence of local geological characteristics on the intensity of ground shaking and damage caused due to an earthquake are found. Data dating back to about 200 years supports this claim. However, it was not until the early 1970s that the local site conditions were included in building codes and provisions. Evidence for the existence of local site effects is quite overwhelming. In addition to theoretical evidence, amplification functions developed from measurements of surface and bedrock motions at the same location, and comparisons of surface motion characteristics from nearby sites with different subsurface conditions, all confirm the effects of local site conditions of earthquake ground motions.

5.2.1 Effect of Topography

Local site effects produce significant amplifications of the ground motion during an earthquake. One of the important local site effects is due to the effect of topography. Irregular topography can substantially affect the amplitude and frequency characteristics of seismic motion. Macro seismic observations of destructive earthquakes often show higher damage intensity at the tops of hills, ridges, and canyons than at lower elevations and on flat areas. Two general types of topography have to be distinguished

- Surface topography, mainly characterized by mountainous features, such as the presence of rock ridges or steep soil slopes,
- Subsurface (or subsoil) topography, either caused by lateral heterogeneities of the subsoil layers or by sharp basin geometry [90, 212].

The curvature of a sediment-filled basin structure in particular can capture body waves and cause some incident body waves to propagate through the alluvium as surface waves resulting in stronger shaking effects and longer duration of strong ground motion [181]. The effect of topography on the amplification has been studied by field experiments. Trifunac and Hudson [361] and Davis and West [75] recorded significant amplifications. But the amplifications recorded were much higher than those predicted by the theoretical models. The main problem involved was development of adequate soil models that can sufficiently represent the uncertainties involved in the geology and geotechnical properties of the area [19, 105]. Aki [6] pointed out that significant correlations exist between the amplification, geological and geotechnical features. Aki [6] also proposed that these complex relationships can be represented by the seismic microzonation maps. Pedersen et al. [264] found that the diffracted waves found at the reference station might explain the amplification of spectral ratios. The quantitative value of the ground motion at any particular site depends upon the source of the seismic waves.

The main effects of topography as pointed out by Brune [50] and Sánchez-Sesma [297, 298] are:

- Amplification of SH waves near the crests of canyons.
- The presence of a shear fundamental resonance in a 3–5 Hz frequency band.
- Dependence on the radiation wave field, angle of incidence, and canyon dimensions.

In contrast to subsoil topography, it is the surface topography that causes more serious effects. According to numerous reconnaissance studies after strong earthquakes, an increase of damage to buildings can be observed on steep slope situations which extend towards the plateau. Recent damaging events like Bingol earthquake of May 1, 2003 in Turkey and Northern Algerian earthquake of May 21, 2003 displayed heavy damage concentrations along the top of steep slopes. According to different scientific groups [6, 19, 105] that deal with instrumental and theoretical investigations of surface topography on ground motion characteristics, the following are stated:

- Mountain tops or ridge crests, and more generally, convex topographies (such as cliff borders), lead to an amplification of seismic ground motion, while valleys or foothills (concave topographies) tend to de-amplify the seismic signals.
- The effects of surface topography are larger on horizontal components than on
- Vertical ones, thus indicating that S motion is more affected by surface topography than P motion.
- The influence of surface topography on ground motion is directly related to the sharpness of topography. According to this theoretical model, amplification of incoming seismic waves increases as the wedge angle becomes sharper.
- Amplification and de-amplification of seismic ground motion on topographic features are both frequency-dependent and band-limited: maximum effects can be observed for wavelengths that approximately agree with the horizontal dimensions of the topographic shape.

The approach adopted by AFPS [2] was to introduce an additional empirical parameter (as in Eq. (5.1)) as the topographical amplification factor in the definition of the design spectrum to account for the topographical effects.

$$\tau = 1 + 0.8(I - i - 0.4)1.0 < \tau < 1.4 \tag{5.1}$$

where I and i are the gradients of the lower and upper slopes, respectively. During the Athens earthquake that occurred in 1999, amplification studies on sites were carried out to find the effect of topography. It was concluded that the average topographic amplification is about 50% for time periods of about 1 s at a site [45]. The effect of slope topography on the seismic ground motion is studied by Bouckovalas and Papadimitriou [46]. The following important observations are to be noted:

- Even a purely horizontal excitation, as a vertically propagating SV wave, results in considerable vertical motion at the ground surface near the slope. This component of ground motion is independent of any vertical excitation induced to the base of the slope by the earthquake itself and, consequently, it has to be superimposed to it. The results of the parametric analyses show that the vertical component of seismic motion may become comparable to the horizontal free-field motion.
- The topography aggravation of the horizontal ground motion, expressed through the peak acceleration ratio $A_h = Z_{ah}/a_{h,ff}$, fluctuates intensely with distance away from the crest of the slope, alternating between amplification ($A_h > 1.0$) and de-amplification ($A < 1.0$) within very short horizontal lengths.

- The horizontal ground motion is de-amplified at the toe of the slope and amplified near the crest. As a result, topography aggravation may be seriously overestimated, when calculated as the peak seismic ground motion at the crest over that at the toe of the slope.

The above two points can be directly associated to reflection of incoming SV waves on the inclined free surface of the slope which leads to reflected P and SV waves impinging obliquely at the free ground surface behind the crest, as well as Rayleigh waves. Topography effects become important for normalized height ratios $H/\lambda > 0.16$ and slope inclinations $i > 17$. If these conditions are met, the peak values of topography aggravation factors for the horizontal and the vertical ground acceleration behind the crest usually vary between $A_{h,\max} = 1.20$–1.50 and $A_{v,\max} = 0.10$–1.10, while free field conditions behind the crest are usually met at a distance $D_{ff} = (2$–$8)H$.

Based on the records obtained from the dense strong motion array in Taipei basin for two earthquakes ($M_L = 6.5$ and 6.57) [197] reported that there were significant differences among the peak ground accelerations, durations, and spectral accelerations in different parts of the basin as well as among the records for two earthquakes. The authors have concluded that for estimating ground motion variations for microzonation of the basin one must be very careful to draw conclusions by using only few seismic events and that it is necessary to collect more data from the array to perform a more detailed microzonation study. The last two decades have witnessed significant progress in estimation of local site effects by different researchers all over the world [43, 56, 299, 362].

The effect of Central Mountain Range in Taipei basin was evaluated by Lee et al. [191]. The Spectral Element Mesh method was adopted to simulate the seismic wave propagation. Realistic topography and complex subsurface topography can be effectively incorporated using SEM. It was concluded that in the event of a shallow earthquake in the area the central mountain range will scatter the surface waves thereby reducing the magnitude of strong ground motion. The occurrence of a deep earthquake, topography scatters the body waves which propagate as body waves resulting in increase of PGA values by over 50%.

Ma et al. [201] studied the effect of San Gabriel mountains which are bounded by the Mojave segment of the San Andreas fault on the north and by the Los Angeles Basin on the south. It was observed that effect of mountain topography reduces the value of ground motion for some areas in the basin. The topography of the mountains scatters the surface waves generated by the rupture on the San Andreas fault, leading to less-efficient excitation of basin-edge generated waves and natural resonances within the Los Angeles Basin.

It is very important to understand the local geotechnical and geological conditions when performing seismic microzonation studies. Site classification can be done based on geological units [233]. But in microzonation even further detailed study of the variations in the geological units is necessary. Wills and Silva [383] suggested and utilized the average shear wave velocity in the upper 30 m as one parameter to characterize the geologic units admitting the importance of other factors such as impedance contrast, three dimensional basin and topographical effects, and source effects such as rupture directivity on ground motion characteristics. Thus the influence of local soil stratification on earthquake characteristics is one of the major factors in evaluating the earthquake forces and thus the structural response. Local soil conditions may amplify or deamplify the earthquake forces in different regions. Sometimes an appropriate simplification may be needed to reflect the complex stratification characteristics in describing the soil amplification phenomena.

5.2.2 *Effect of Groundwater*

The relation between groundwater and earthquakes is not new and is well documented. The Chinese noticed centuries ago that water levels in wells can vary in association with earthquake activity, and used this behavior with some success to anticipate earthquakes. More recent research in the United States has attempted to use monitoring of groundwater levels in wells to help predict earthquake activity [226]. The 1983 Borah Peak earthquake in southeastern Idaho caused groundwater located near its epicenter to erupt as much as 25 ft into the air [71]. However, within the past 50 year researchers have noted that groundwater can play a direct role in earthquake occurrence and earthquake related damage. In addition, it is known that earthquakes such as the 2004 Indonesian quake can cause measurable changes on groundwater levels in places thousands of miles from the epicenter [368]. Research on earthquake mechanisms indicates that groundwater likely plays a significant role in many large earthquakes. Furthermore, groundwater can magnify the damaging effects of ground surface.

Groundwater can also play a significant role in how earthquakes affect the ground surface when an earthquake occurs. The most well-known effect is liquefaction. During the shaking cause by an earthquake, certain types of fluid saturated sediments can lose their structure and become liquefied. These phenomena led to increased structural damage. The geologic and hydrologic factors that affect liquefaction susceptibility are the age and the type of sedimentary deposits, the looseness of cohesions less sediments, and the depth to the groundwater table. The liquefaction is mostly limited to water-saturated, cohesions less, granular sediments at depths less than 15 m. Noack and Fäh [240] gave weight according to the depth of water table (Table 5.1 more weight more damage, less weight less damage).

Table 5.1 Weightage factor for assessing liquefaction susceptibility

Depth of water table	Weight
10–20 m	2
3–10 m	3
1–3 m	4

During an earthquake, base rock movements generate shear waves that propagate through overlying soils. Liquefaction results when these shear waves, passing though saturated sand layers, distort the granular structure and cause loosely packed grains to collapse. This densification causes an increase in pore pressure if drainage cannot occur. If these pore pressures exceed roughly sixty percent (60%) of the soil's effective stress, large settlements and translational deformations can occur.

Groundwater level influences the ground response significantly and cannot be neglected for site effect analyses. In the presence of confined aquifers liquefaction can take place in the subsurface. This results in attenuation of the propagation of shears waves and can reduce the ground shaking. In the presence of groundwater the shear wave cannot reach the surface as they can't pass through the water, because water offers no shear resistance.

5.2.3 Effect of Bedrock

In many earthquakes, the local geology and soil condition have had profound influence on site response. The term local is somewhat vague one generally meaning local compared to the total terrain transverse between the earthquake source and the site. On the assumption that the gross bedrock vibration will be similar at two adjacent sites. Local differences in geology and soil properties will cause different surface ground motions at the two sites. Factors influencing the local modification to the underlying motion are the topography and the nature of the bedrock and the nature and geometry of the depositional soils. Thus, the term local may involve a depth of a kilometer or more, and an area within a horizontal distance of several kilometers from the site. Soil conditions and local geological features affecting site response are numerous, and are now discussed as below:

- The greater the horizontal extent of the softer soils, the less the boundary effects of the bedrock on the site response.
- The depth of soil overlaying bedrock affects the dynamic response, the natural period of vibration of the ground increasing depth.

This helps to determine the frequency of the wave amplified or filtered out by the soils and is also related to the amount of soil structure interaction that will occur in an earthquake. The Mexico earthquake of 1957 and 1985 witnessed extensive damage to long-period structures in the former lake bed area of Mexico city where the flexible lacustrine deposits caused greater amplification of long period waves.

- The slope of the bedding planes of the soils overlying bedrock obviously affects the dynamic response, but it is less easy to deal rigorously with non-horizontal strata.
- Dynamic response, but it is less easy to deal rigorously with non-horizontal strata.
- Changes of soil type horizontal across a site affect the response locally within that site, and may profoundly affect the safety of a structure straddling the two soil types.

- The topography of both the bedrock and the deposited soils has various effects on the incoming seismic waves, such as reflection, refraction, focusing, and scattering.

5.3 Methods for Estimating Local Site Effects

As the local site effects play a crucial role in aggravating the damage potential of an earthquake, it must be accounted in the earthquake resistant design. There are many methods available for the estimation of site effects which can be broadly categorized as (1) Empirical methods, (2) Experimental methods, and (3) Numerical studies.

5.3.1 Empirical Methods

Many researchers have observed significant dependency of site amplification on the local surface geology based on the strong motion records of many earthquake events. Based on earthquake records for each geological formation, many empirical relations between surface geology and various ground motion parameters have been developed which can be applied in another region having same geology. These empirical relations will give only an approximate value of site amplification and hence is good for first level assessment. Intensity based empirical correlations proposed by Medvedev [214] (Table 5.2), Evernden and Thomson [87], Kagami et al. [158] and Astroza and Monge [15] are the major ones which were developed using earthquake records in Asia, California, Japan, and Chile, respectively.

Table 5.2 Intensity increment for each geological unit [87]

Lithology	MM scale
Granitic and metamorphic rock	0
Paleozoic rocks	0.4
Early mesozoic rocks	0.8
Cretaceous to eocene rocks	1.2
Undivided tertiary rocks	1.3
Oligocene to middle Pliocene rocks	1.5
Pliocene-Pleistocene rocks	2.0
Tertiary volcanic rocks	0.3
Quaternary volcanic rocks	0.3
Alluvium (water table < 30 ft)	3.0
Alluvium (30 ft < water table < 100 ft)	2.0
Alluvium (water table > 100 ft)	1.5

Table 5.3 Correlations between surface geology and relative amplification [144]

Previous works	Geology	Relative amplification
Borcherdt and Gibbs [42]	Bay mud	<11.2
	Alluvium	3.9
	Santa Clara formation	2.7
	Great valley sequence	2.3
	Franciscan formation	1.6
	Granite	1.0
Shima [324]	Peat	1.6
	Humas soil	1.4
	Clay	>1.3
	Loam	1.0
	Sand	0.9
Midorikawa [216]	Holocene	3.0
	Pleistocene	2.1
	Quaternary volcanic rock	1.6
	Miocene	1.5
	Pre-tertiary	1.0

Many researchers [42, 216, 324] have given the relative amplification factors for different geological and soil conditions (Table 5.3) based on analytical studies. Other researchers [44, 157, 216] have proposed empirical relations relating the average shear wave velocity of surficial layers to the relative amplification. Topographical effects on the earthquake ground motion characteristics have been studied by Geli et al. [105], Faccioli [88] and Chávez-García et al. [65].

Many researchers [39, 274, 345] have proposed various empirical relations for various site classes as per NEHRP scheme. These site coefficients and empirical equations for estimating PHA and S_a for each NEHRP site classes were developed using regression analysis. The assessment of site effects based on local geology is only approximate and may not incorporate deep basin effects or topographic effects. Thus this methodology provides a less reliable estimate of site amplification and hence much suitable for a first level study for a large landmass.

The most important geotechnical parameters, which can be used to estimate the amplification factors are average shear wave velocity and SPT 'N' value. In the early days measurement of shear wave velocity using cross hole tests were expensive whereas the SPT 'N' values were available abundantly. Shima [324] found that the analytically calculated amplification factor is linearly related with the ratio of shear wave velocity of the surface layer to that of bedrock. When the bedrock shear wave velocity is found to be relatively constant over a wide area, the relative amplification in each locality can be obtained from the shear wave velocity of the surface layer. Various researchers [44, 157, 216] proposed relations between the average shear wave velocity of surficial layers and the relative amplification as shown in Table 5.4.

Table 5.4 Correlations between average shear wave velocity and relative amplification [144]

Author	Proposed equation	
Joyner and Fumal [157]	$RA = 23(V_s)^{-0.45}$	
Midorikawa [216]	$RA = 68(V_{s30})^{-0.6}$	$(V_{s30} < 1100\,m/s)$
	$RA = 1.0$	$(V_{s30} < 1100\,m/s)$
Borchedt et al. [44]	$AHSA = 700/V_{s30}$	(for weak motion)
	$AHSA = 600/V_{s30}$	(for strong motion)

5.3.2 *Experimental Methods*

These methods are based on different kinds of data obtained from microtremor measurements, weak seismicity survey, and strong motion data. Detailed description for estimating site effects using the above methods is given below. Kanai[162] have presented the use of the microtremor method for site response evaluations. Microtremors are vibrations, having comparatively small periods (less than 1 s). Udwadia and Trifunac[366] compared the site amplification obtained from microtremor method and strong motion records. Nakamura [229] has proposed H/V ratio technique to predict the amplification factor for a site.

(1) Microtremor Data

The site effects are often expressed by the amplification factor and resonance or fundamental frequency. Usually there are various vibrations in the ground which are caused by natural or ambient noise like wind, sea waves, traffic, industrial machinery, etc. The range of vibration frequencies of ambient noises is from 0.1 to 10 Hz (i.e., 10–0.1 s period). The vibrations that have comparatively small periods of less than 1 s are called microtremors and those that have a larger period range is called microseisms. Due to close relation between spectral features of microtremors and site's geological conditions, these small vibrations are very useful in earthquake geotechnical engineering. Kanai [162] have explained a theoretical interpretation and practical engineering application of microtremors as a convenient tool for evaluating frequency properties of surface ground. The use of microtremor method for site response evaluations has been restricted to Japan with many controversies in USA and Europe. Applications of this method include site response analysis, natural frequency of structures, etc. However, the real generation and nature of microtremors have not yet been established.

Kanai [161] proposed a method to classify the ground into four categories, which is used by the Japan Building Code. This classification is based on the detailed comparison of microtremor results and ground conditions. A general wave pattern and period-frequency (PF) diagrams for four types are shown in Fig. 5.2. It is clear that a stiff ground/rock is inclined to have low predominant period and soft sediments have high predominant period. Kanai [162] proposed two methods based

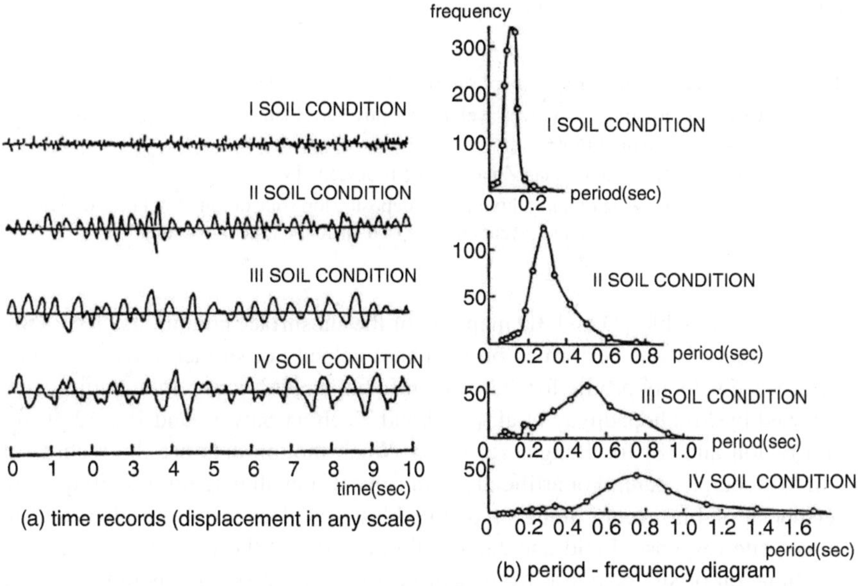

Fig. 5.2 Classification of microtremor types [163]. (**a**) Time records (displacement in any scale). (**b**) Period–frequency diagram

on microtremor records. One is based on the largest period and the mean period, the other based on the largest amplitude in microns and the predominant period. This can be used to identify site categories associated with various levels of seismic damage due to strong shaking. Table 5.5 gives the detailed description of soil type in each category. However it was found that the H/V technique is not well correlated with the S wave amplification at the site's resonant frequency [179, 243]. Though the H/V method is good to map the fundamental resonance, it cannot give a good estimation of the amplification value at a site [115]. It is highly sensitive to some parameters like Poisson's ratio near the surface. A one-to-one average correlation is developed by Konno and Ohmachi [179] on the basis of a comparison between observed H/V peaks and numerical estimates of 1D transfer functions. Bard et al. [20] compared the amplifications derived from earthquake records and H/V peak amplitudes for more than 30 sites and demonstrated that the latter is always smaller than the former. This Nakamura method provides a lower bound estimate to the actual amplification which is proved by large experimental data.

Aki [3] has measured phase velocities of surface waves by analyzing the spatial correlation of the microtremors. Then, the P and S wave velocity profiles can be derived by inversion which makes it possible to estimate the site response by 1D modelling. Several studies in Italy [69, 204], Japan [220] and Israel [108] illustrate the practical use of this technique with other methods used in geotechnical engineering in estimating velocity profiles. In particular, its coupling with the Nakamura technique (array measurements at few sites and H/V ratios at many sites)

Table 5.5 Microzones for Japan Building Code (Kanai [162])

Zones	Soil description
I	Ground consisting of rock, hard sandy soils or gravely deposits
II	Ground consisting of sandy gravel, hard sandy clay, loam or alluvial gravel with thickness of 5 m or more
III	Standard ground other than Zone I, Zone II or Zone IV
IV	Ground consisting of soft alluvial delta deposits, top soils or mud thickness of 3 m or more where less than 30 years has elapsed since the time of reclamation

may lead to a reliable 2D and 3D mapping of the subsurface conditions [108, 356]. Inversion analysis was performed on the phase velocity of surface waves to obtain a shear wave velocity profile for that site. In India, major microtremor studies were conducted by Mukhopadhyay et al. [227] and Neelima Satyam and Rao [239] for Delhi region and [8] for Bangalore region. Weak motion data are the records of small to moderate, natural or artificial seismic events (small magnitude earthquakes, aftershocks of big events, mine or quarry blasts, nuclear tests) are also used to estimate site response. Field and Jacob [92] quoted that the greatest challenge in the estimation of site response from such instrumental recordings is removing the source and path effects. Many researchers have used strong motion records of past earthquake in order to predict site amplification [24, 38, 172].

(2) Weak Motion Data

Weak motion data are the records from small to moderate, natural or artificial seismic events (small magnitude earthquakes, after shocks of big events, mine or quarry blasts, nuclear tests). Such data can be recorded by digital, high sensitive instruments identical to those used by seismologists for microseismicity and seismotectonics studies. Field and Jacob [92] quoted that the greatest challenge in the estimation of site response from such instrumental recordings is removing the source and path effects. Two techniques are developed depending on whether or not they need a reference site with respect to which the particular effects at other sites are estimated.

(a) Reference Site Technique

The three important factors which will affect the ground motion are the source, path, and the site characteristics. The identification and removal of these effects is the greatest challenge in evaluating the site response. The simplest method to evaluate the site response is to divide the response spectrum obtained at the site with that of the bed rock (reference site). If the recording in the rock is at a close distance to the soil site, then the three governing factors, which will affect the ground motion, will be the same for both the soil site and the rock. However when the reference site and the site under consideration (soil site) are not near by, then the influence of source, path, and the site characteristics will be different for these two sites. Moreover the geometric spreading of the seismic waves will also need to be accounted [40, 41, 43, 122]. To evaluate the source and site terms simultaneously, a generalized inversion scheme developed by Andrews [12] can be applied. In the generalized inversion technique, a relatively large dataset can be used [345] but the non-linear response of sedimentary deposits cannot be predicted accurately when the weak motion data are dominating the input data.

(b) Non-Reference Site Technique

In practice, adequate reference sites are not always available. For this reason, different methods without reference sites have been developed. It consists of taking the spectral ratio between the horizontal and the vertical components of the shear wave part. This technique is a combination of [188] receiver function method for determining the velocity structure of the crust from the horizontal to vertical spectral ratio (HVSR) of teleseismic P-waves and the [229] method. It was first applied to the S-wave portion of the earthquake recordings obtained at three different sites in Mexico city by Lermo and Chávez-García [192]. The same technique has been applied on various sets of weak and strong motion data [32, 65, 187, 284, 353, 402] from which it is concluded that the HVSR shape exhibits a very good experimental stability, it is well correlated with surface geology and less sensitive and comparisons with theoretical 1D computations, the absolute level of HVSR depends on the type of incident waves. Also, the determination of the absolute level of amplification from only HVSR is not straightforward. Field and Jacob [92] applied this technique and found that the method reproduces well the shape of the site response, but underestimates the amplification level. They also found very different results when applying this technique to the P-wave part of the recordings.

(3) Strong Motion Data

The development of strong motion arrays makes it possible to evaluate site effects in mega cities like Los Angeles, Tokyo, Taipei, Mexico city using the strong motion data. While using this method, even the non-linear site effects are included in the recordings. Recent studies show that there is a fairly good agreement between the old and new techniques. With the increase in strong motion data, it is possible to determine soil amplification and site response maps based on ground motion records [17, 24, 38, 121, 352]. Khoubbi-Al and Adams [172] estimated soil amplification in Ottawa, Canada using strong motion records.

5.3.3 Numerical Approach

Numerical methods of estimating site amplification mainly include ground response analysis. Ground response analysis techniques are often categorized based on the dimensionality of the problem and the complexity of the soil model employed [181]. On the basis of dimensionality there are one-, two- and three-dimensional ground response analysis. Similarly based on the soil model used, the three can be further grouped into linear, equivalent linear, and non-linear.

(1) 1D Ground Response Analysis

One-dimensional (1D) ground response analysis assumes that all soil layers are horizontal and soil response is mainly due to the vertically propagating SH-waves only [181]. Evaluation of transfer functions is the key to any 1D analysis. Transfer function relates the base motion to the surface motion in the Fourier domain. In a linear 1D analysis, dynamic soil properties such as the shear modulus (G) and damping ratio (β) are assumed to be consistent with the level of strain while evaluating the transfer function. However the equivalent linear analysis proposed by Seed and Idriss [307] considers the dependency of shear modulus (G) and damping ratio (β) with the level of strain induced. The non-linear analysis is carried out by integrating the equation of motion in small time intervals using any non-linear stress-strain models or advance constitutive models. Several computer programs such as SHAKE [132, 302] and DYNEQ [397], DEEPSOIL [123] are available in order to implement this equivalent linear approach. Similarly there are many computer codes available for carrying out non-linear 1D, ground response analysis DESRA-2 [190], DYNA 1D [270], DEEPSOIL [123].

The 1D equivalent linear approach consists of the following steps: (1) The input earthquake acceleration time history is transformed into frequency domain using a Fourier transfer algorithm. (2) The in-situ soil state is captured by assigning a low strain shear modulus ($G_{\max}$) and damping ratio ($\beta_{\min}$) each soil layer. Appropriate modulus degradation and damping curves are assigned to each layer to capture the strain dependency of shear modulus and damping ratio. (3) For the input motion (in the frequency domain), the response of each soil layer is computed considering $G_{\max}$ and $\beta_{\min}$, and shear strain time histories for each soil layer is generated. (4) Effective shear strain (γ_{eff}) for each layer is then calculated which is approximately 65% of the peak strain amplitude ($\gamma_{\max}$) occurred in the strain time history for corresponding layer. (5) For the effective shear strain (γ_{eff}) value, corresponding shear modulus and damping ratio are evaluated from the modulus reduction and damping curve which are assigned to each soil layer. (6) Step 3 is repeated with a new strain dependent value of shear modulus and damping ratio. The iteration procedure is repeated till concordant values of shear modulus and damping ratio from two successive steps are obtained. In this study, 1D equivalent linear ground response analysis for the study area is carried out using SHAKE 2000 [254] and surface motion, response spectra, predominant frequency, and spectral ratio are evaluated. In a general 1D equivalent linear analysis, the geo-materials (soil or rock) are modelled using the Kelvin–Voigt model and the wave propagation through these solids can be mathematically written as per Eq. (5.2) as given by Kramer [181].

$$\rho \frac{\partial^2 u}{\partial t^2} = G \frac{\partial^2 u}{\partial z^2} + \eta \frac{\partial^3 u}{\partial z^2 \partial t} \tag{5.2}$$

where u is displacement, t is time and z is the vertical direction along which wave propagates. The general solution for the above equation is given in Eq. (5.3).

$$u(z, t) = Ae^{i(\omega t + k^* z)} + Be^{i(\omega t - k^* z)} \tag{5.3}$$

Here, A and B are the amplitudes of the incident and the reflected waves, k^* is the complex wave number $k^* = \sqrt{\frac{\rho}{G^*}\omega^2}$ and $G^* = G(1 - 2\beta^2 + i2\beta\sqrt{1 - \beta^2})$ is the frequency independent complex shear modulus. 1D ground response study analyse the vertical propagation of shear waves through soil layers lying on an elastic rock. The elastic bedrock is assumed to be extended to infinite depth. Hence Eqs. (5.4) and (5.5) (where subscripts s and r refers to soil and rock respectively) present the response of soil and elastic rock due to the upward propagation of shear waves.

$$u_s(Z_s, t) = A_s e^{i(\omega t + k^* z_s)} + B_s e^{i(\omega t - k^* z_s)} \tag{5.4}$$

$$u_r(Z_r, t) = A_r e^{i(\omega t + k^* z_r)} + B_r e^{i(\omega t - k^* z_r)} \tag{5.5}$$

Certain boundary conditions are then applied to reduce the redundancies. These conditions are (1) at free surface, the shear stress ($\tau_s = 0$) and shear strains are zero ($\gamma_s = 0$), (2) displacements and stresses at soil–rock interface are equal and continuous (i.e. $u_s(z_s = H) = u_r(z_r = 0)$, $\tau_s(z_s = H) = \tau_r(z_r = 0)$), (3) compatibility of displacements and stress along the layer boundaries, i.e. $u_m(z_m = H_m) = u_{m+1}(z_{m+1} = 0)$, $\tau_m(z_m = H_m) = \tau_{m+1}(z_{m+1} = 0)$ where m and $m + 1$ are successive soil layers and as shown in Fig. 5.3.

From the compatibility conditions, amplitudes of incident waves and reflected waves in the $(m + 1)$th (A_{m+1}, B_{m+1}) layer can be expressed in terms of amplitudes in mth layer (A_m, B_m) as given in Eqs. (5.6) and (5.7).

$$A_{m+1} = \frac{1}{2} A_m (1 + \alpha_m) e^{ik_m^* h_m} + \frac{1}{2} A_m (1 - \alpha_m) e^{-ik_m^* h_m} \tag{5.6}$$

$$B_{m+1} = \frac{1}{2} B_m (1 + \alpha_m) e^{ik_m^* h_m} + \frac{1}{2} B_m (1 - \alpha_m) e^{-ik_m^* h_m} \tag{5.7}$$

where $\alpha_m = \frac{k_m^* G_m^*}{k_{m+1}^* G_{m+1}^*}$ is the complex impedance ratio at the boundary between the mth layer and $(m + 1)$th layer [181]. For the free surface condition, $A_1 = B_1$ can be derived. Starting with the surface layer, repeated use of recursion formulas (in Eqs. (5.6) and (5.7)) will result in Eqs. (5.8) and (5.9) relating amplitudes in mth layer with those in the surface layer.

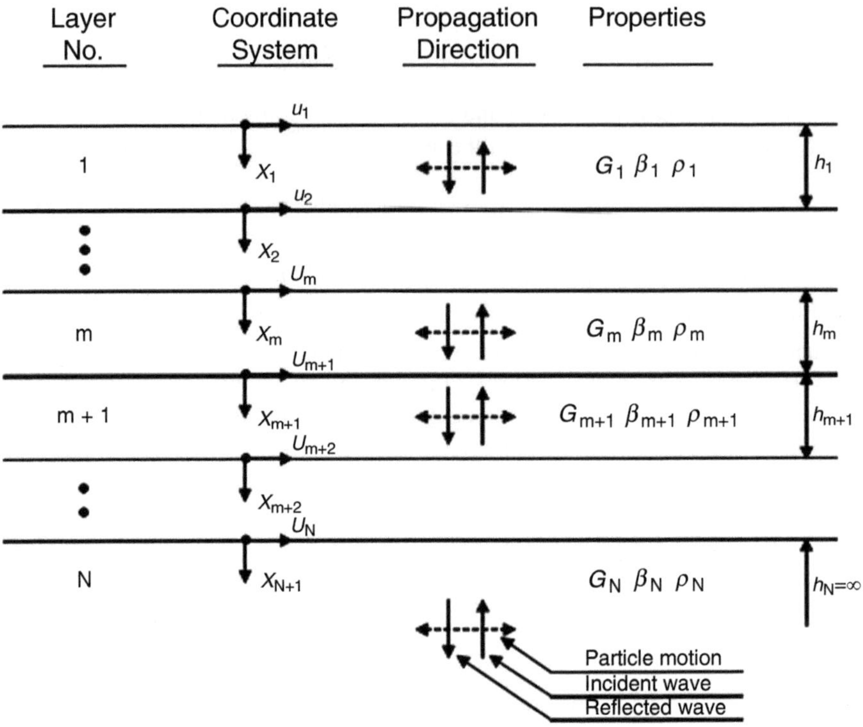

Fig. 5.3 Representations of layered soil (after [302])

$$A_m = a_m(\omega)A_1 \tag{5.8}$$

$$B_m = b_m(\omega)B_1 \tag{5.9}$$

Now the frequency dependent transfer function relating the displacement at ith and jth layer $F_{ij}(\omega)$ can be expressed as per Eq. (5.10).

$$F_{ij}(\omega) = \frac{u_i}{u_j} = \frac{a_i(\omega) + b_i(\omega)}{a_j(\omega) + b_j(\omega)} \tag{5.10}$$

Once amplitudes A and B of all layers are obtained, acceleration and strain histories of any layer can be computed using Eqs. (5.11) and (5.12), respectively.

$$\frac{\partial^2 u}{\partial t^2} = -\omega^2\left(Ae^{i(\omega t + k^* z)} + Be^{i(\omega t - k^* z)}\right) \tag{5.11}$$

$$\frac{\partial u}{\partial z} = -ik\left(Ae^{i(\omega t + k^* z)} - Be^{i(\omega t - k^* z)}\right) \tag{5.12}$$

The effective shear strain (γ_{eff}) for equivalent linear analysis is calculated from the maximum shear strain ($\gamma_{\max}$) value in the shear strain time histories of each layer. The effective shear strain is given as $\gamma_{eff} = R_{\gamma}\gamma_{\max}$ where the stress reduction factor R_{γ} is a function of earthquake magnitude. However it is found that R_{γ} varies from 0.5 to 0.7 and is not sensitive on response, hence in SHAKE 2000 it is taken as 0.65.

(2) 2D and 3D Ground Response Analysis

A two-dimensional (2D) analysis is normally carried out for the sites having sloping/irregular ground surface or locations where deep basin formations are present. The curvature of a soft alluvium deposited basin can trap body waves and cause them to propagate as surface waves. Hence the shaking produced will be of higher intensity and of longer duration when compared with those predicted using 1D analysis [181]. Three-dimensional (3D) analysis is the most rigorous analysis which is generally carried out to understand the response of three-dimensional structures. 2D and 3D analyses are generally carried out using dynamic finite analysis. 2D analyses can also be performed using a simplified methodology called shear beam approach [103]. Numerical modelling software like PLAXIS, FLAC, QUAKE/W, etc. can be used for modelling two-dimensional case.

The ground response analysis requires geotechnical and geophysical test data as input parameters in order to compute soil response. The 2D and 3D ground response analyses based on finite element tools require lot of computational power. Hence this method is most suitable to micro to meso-level studies.

5.4 Evaluation of Site Amplification at Micro-Level: A Case Study of Kalpakkam NPP Site

Ground response studies at the micro-level for the Kalpakkam NPP site are carried out by James et al. [153] to obtain ground surface motion, surface level response spectrum, and predominant frequency of surface motion. As the nuclear power plant structures are critical, rigorous studies are needed to be carried out for the estimation of site amplification. The amplifying effect that the site can have on earthquake motions at bedrock can reasonably be quantified by conducting ground response analyses. In the present study, 1D equivalent linear ground response analysis for the Kalpakkam NPP site was carried out using SHAKE 2000 [254].

SHAKE 2000 is a Windows based computer program for ground response analysis for the evaluation of the earthquake effects on soil deposits. SHAKE 2000 [254] is an upgraded version of SHAKE program developed by Schnabel et al. [302] at the University of California Berkeley. SHAKE 2000 is a software package that integrates ShakEdit and SHAKE. ShakEdit was originally developed as a 16-bit,

Windows 3.1 application that provides a graphical interface for SHAKE. SHAKE computes the response of the soil layers subjected to earthquake excitation at the base in the frequency domain. The non-linear behavior of the soil is approximated using an equivalent linear approach incorporating an iterative procedure to obtain the strain dependent values for modulus and damping in each of the soil layers.

5.4.1 Generation of Synthetic Earthquake Strong Motion

One-dimension ground response analysis requires an earthquake acceleration time history data of the site, which has to be given as input in the program. As there are no suitable earthquake records available for the study area available in any major database, a synthetic acceleration time history was generated by James et al. [153] for the site response analysis. There are many methodologies available for the generation of synthetic acceleration time history like empirical green function technique [120, 137] and stochastic simulation technique [35]. The above two methods require the knowledge of complex seismological parameters and hence in this work, a simple method is used, which synthesizes the artificial ground motion compatible with the target acceleration response spectrum. A MATLAB code was developed to generate by James et al. [153] artificial earthquake time history based on the methodology proposed by Zhao and Zhang [403]. In this method a target response spectrum was first selected. The response spectrum obtained from deterministic hazard analysis of the study area (Fig. 5.4) was used as target response

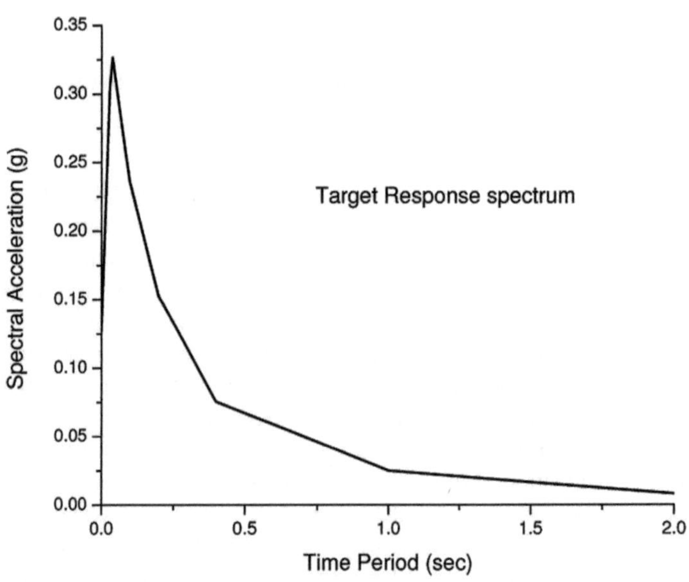

Fig. 5.4 Target response spectrum obtained from DSHA (after [153])

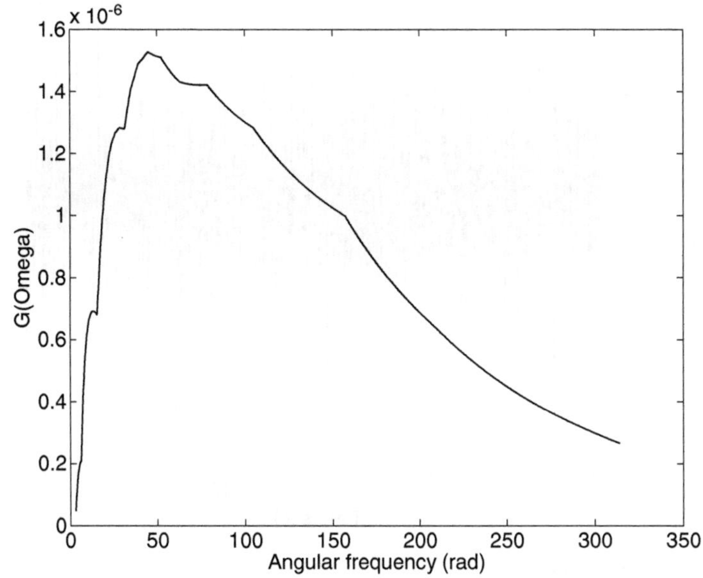

Fig. 5.5 Power density spectrum (after [153])

spectra. The target response spectrum was transformed into the corresponding power spectrum (Fig. 5.5) by the following approximate transfer formula [169] described by Eq. (5.13).

$$G(\omega) = \left[\frac{\beta}{\pi\omega}\right]\left[\frac{S_T^2(\omega,\beta)}{\ln(\omega T) - \ln[-\pi\ln(l-r)]}\right] \tag{5.13}$$

where $G(\omega)$ is the power spectrum, $S_T(\omega, \beta)$ is the target response spectrum, β is the damping ratio, T is the duration of seismic ground motion, and r is the probability of exceeding the target response spectrum. A stationary signal (Fig. 5.6) of duration ($T = 15$ s) was generated as per Eq. (5.14).

$$X(t) = \sum_{n=0}^{N-1} A(\omega_n)\cos[\omega_n t + \phi(\omega_n)] \tag{5.14}$$

where $A(\omega_n) = [4G(\omega_n)\Delta\omega]^{0.5}$, $\Delta\omega = (2\pi/T)$ is the frequency-domain sampling distance, and the initial phase $\phi(\omega_n)$ is a random variable with uniform distribution within the interval of $[0, 2\pi]$. The stationary time history $X(t)$ was then multiplied with a prescribed non-stationary intensity envelop $E(t)$ in order to transform it into non-stationary acceleration time history (Eq. (5.15)).

$$a_g(t) = E(t)X(t) \tag{5.15}$$

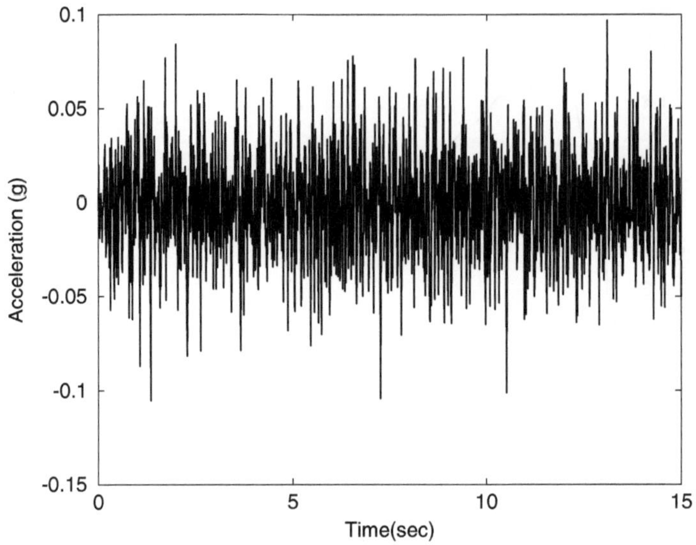

Fig. 5.6 Stationary signals

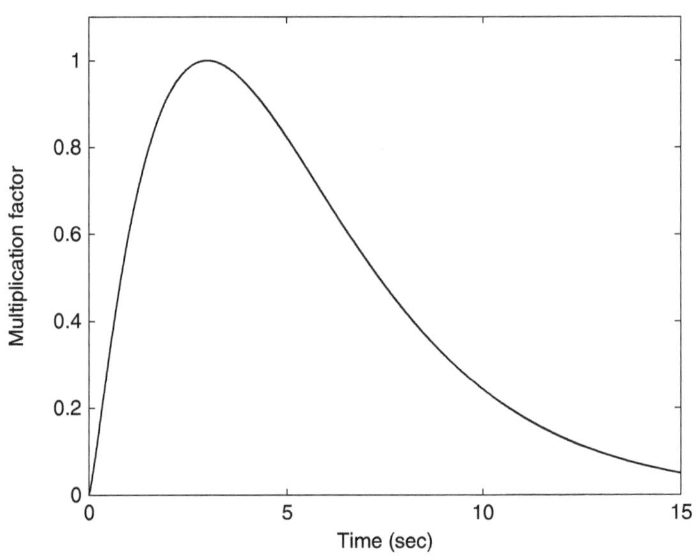

Fig. 5.7 Envelop function

where $E(t)$ is the envelope function (Fig. 5.7) given by Boore [34] and $a_g(t)$ is non-stationary acceleration time history (Fig. 5.8) as in Eq. (5.16). Same envelop function was employed by Kumar [184].

$$E(t) = at^b e^{-ct} \tag{5.16}$$

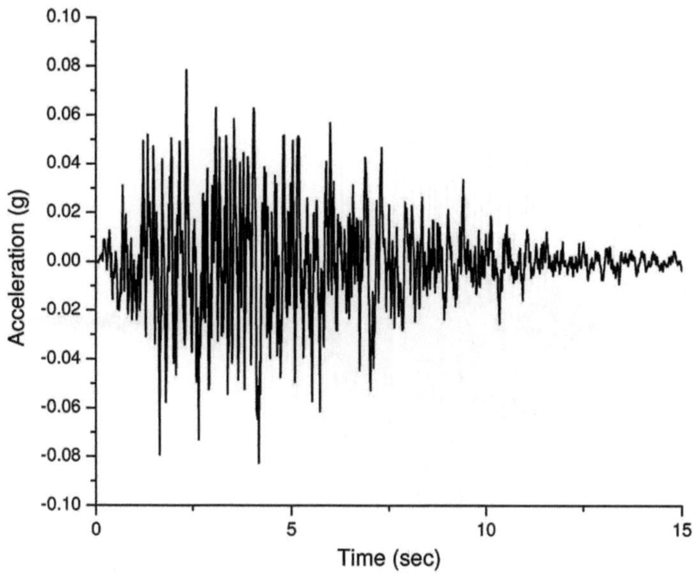

Fig. 5.8 Non-stationary time signal

where $a = \left[\frac{e}{\varepsilon T_d}\right]^b$ $b = \left[\frac{-\varepsilon \ln \beta}{1+\varepsilon(\ln \varepsilon - 1)}\right]$ and $c = \left[\frac{c}{\varepsilon T_d}\right]$ and in this work, ε was taken as 0.2 and β was taken as 0.05 and T_d is the total duration of ground motion (assumed as 15 s). The transformation of response spectrum to power spectrum using Eq. (5.13) is only approximate, so the response spectra $S_a(\omega, \beta)$ obtained from the generated time history (in step 2) rarely matches with the original or target response spectra $S_T(\omega, \beta)$. Hence an iterative procedure was employed to explicitly match the observed response spectra with target response spectra so as to obtain a compatible time history. The commonly-used iterative method modifies the Fourier amplitude spectrum in Eq. (5.17).

$$A^{i+1}(\omega_j) = \frac{S_T(\omega, \beta)}{S_a(\omega, \beta)} A^i(\omega_j) \tag{5.17}$$

where $A^i(\omega_j)$and $A^{i+1}(\omega_j)$ are the results of the ith and the $(i+1)$th iteration, respectively, and $S_T(\omega, \beta)$ and $S_a(\omega, \beta)$ are the target spectrum and the calculated spectrum at the matching frequency of ω_j. A number of iterations are performed and the observed response spectrum is matched to the target. The final acceleration time history for the site is presented in Fig. 5.9. Plot showing the target response spectrum and observed response spectrum is given in Fig. 5.10.

This methodology is very effective as it that modifies the Fourier amplitude spectrum of time history to fit the target acceleration spectrum. By superimposition of narrow-band time histories in the time domain, this method improves the fitting precision of the artificial ground motion to the target response spectrum. The

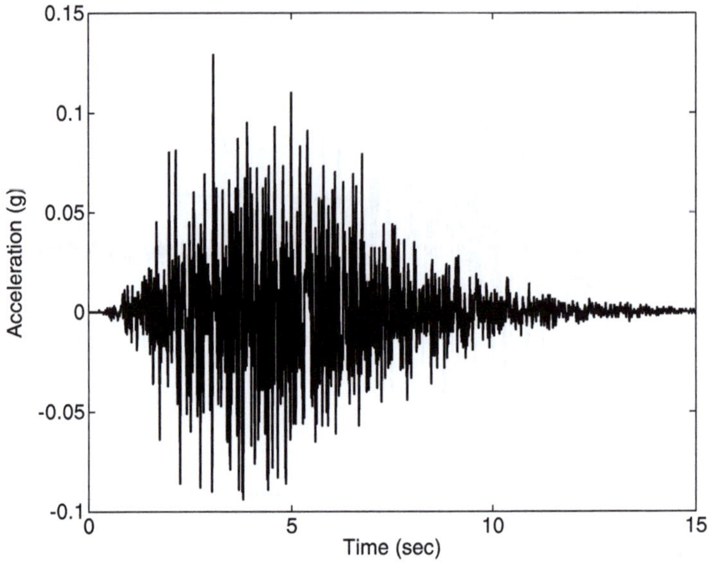

Fig. 5.9 Synthetic strong motion for the study area, compatible with TRS (after [153])

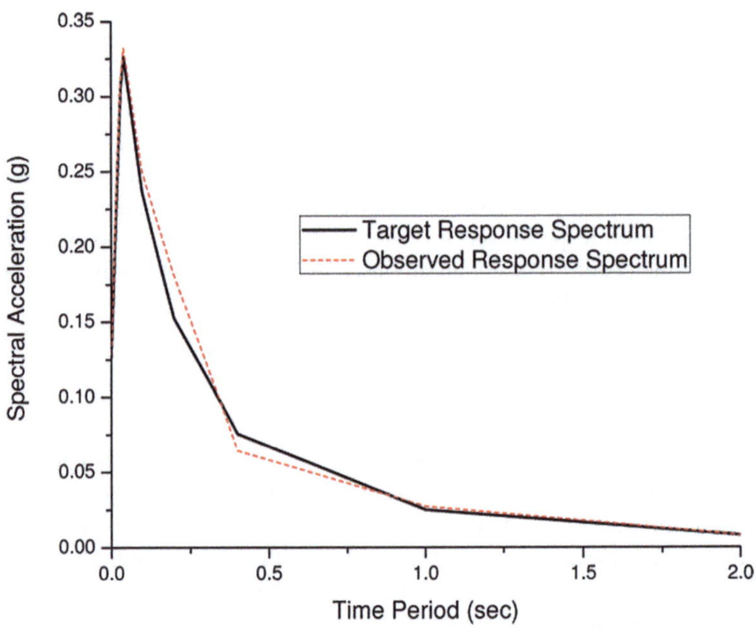

Fig. 5.10 Compatibility between observed and target response spectrum (after [153])

Table 5.6 Details of typical subsurface profile in the study area (after [153])

Strata name	In-situ density (kN/m^3)	Particle size C	M	S	G	N-value observed	Index properties W_L	IP
Sand	17–19	0	2	98	0	10–40	–	–
Soft clay	17–20	64	36	0	0	2–20	91	65
Filled up soil	20–22	5	20	75	0	>50	35	20
Residual soil	19–23	0	4	57	39	>50	–	–
Weathered rock	25–27	–	–	–	–	>100 (RQD 0–50%)	–	–
Rock	27	–	–	–	–	>100 (RQD > 100%)	–	–

synthetic ground motion generated (in Fig. 5.9) has a peak acceleration of 0.129 g at 3.07 s, peak velocity of 2.630 cm/s at 5.48 s, and peak displacement of 0.435 cm at 5.39 s.

5.4.2 *Modelling the Soil Profile*

Stratifying and assigning appropriate modulus reduction and damping model for the soil overburden is an important step in ground response analysis. Stratum identification was done by James et al. [153], based on the borelog data and shear wave velocity data. Field test data revealed the existence of five types of soil layers which were identified mainly as sand, soft silty clay, residual soil (gravelly Sand), weathered rock and bedrock (Charnockite rock type). Typical characteristics of each of the soil samples are described in Table 5.6. Borelog data also reveal the presence of dense filled up soil between BH-5 and BH-6 boreholes (Fig. 4.9), having brownish coloration which is a mixture of clay and sand.

After the characterization of soil stratum the important task was to assign appropriate modulus reduction curve and damping curve. SHAKE 2000 provides different types of modulus reduction and damping curves proposed by many researchers for different types of geo-materials. Details of modulus reduction and damping curves used for each stratum are given in Figs. 5.11 and 5.12 and Table 5.7. Modulus reduction and damping curve proposed by Vucetic and Dobry [377] for soil having an overconsolidation ratio up to 15 is used for modelling the dynamic behavior of the filled up soil.

5.4.3 *Ground Response Analysis*

One-dimensional equivalent ground response analysis was carried out using SHAKE software packages. Analysis was carried out at eight major locations in the Kalpakkam NPP site and the results are briefly presented in Table 5.8. The spatial

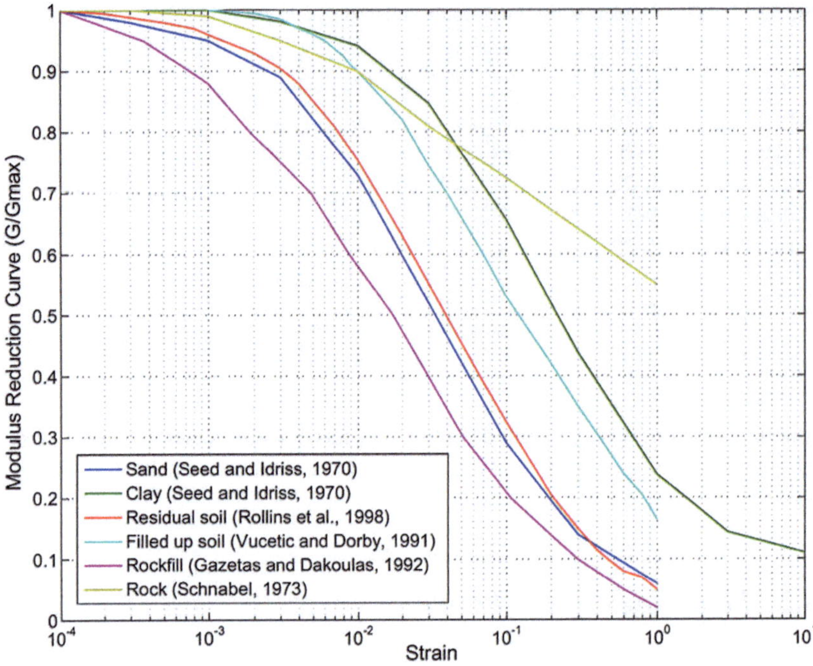

Fig. 5.11 Modulus reduction curves used for various soil layers in the study area (after [150])

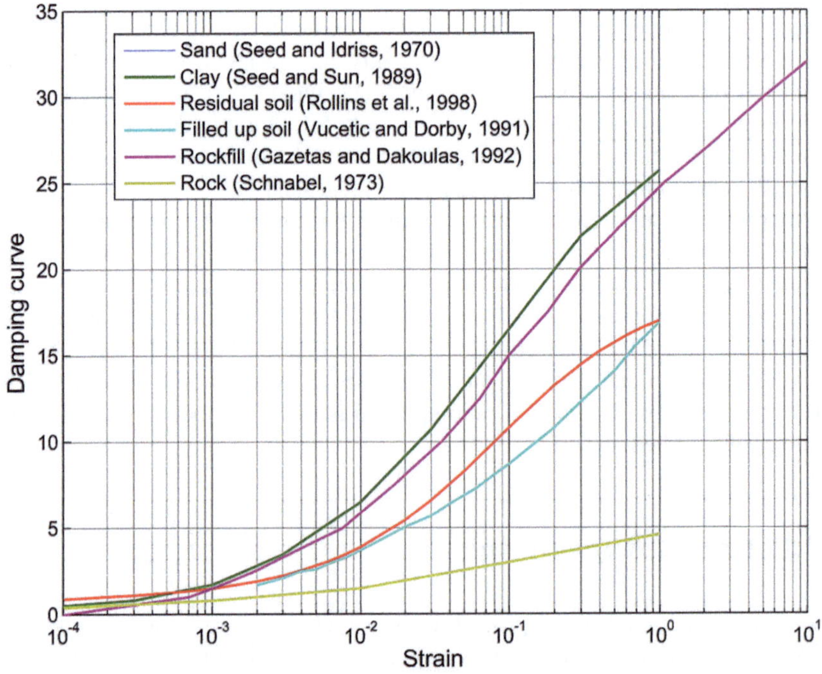

Fig. 5.12 Damping curves used for various soil layers in the study area (after [150])

Table 5.7 Dynamic properties assigned to each soil type (after [153])

Sl. no.	Soil type	Modulus reduction curve	Damping curve
1	Sand	Sand Ave [308]	Sand Ave [308]
2	Clay	Clay upper range [308]	Clay upper range [312]
3	Residual soil	Gravel mean [292]	Gravel mean [292]
4	Weathered rock	Rockfill [104]	Rockfill [104]
5	Hard rock	Rock [301]	Rock [301]

Table 5.8 Results of 1D equivalent ground response analysis (after [153])

Location description	PHA (g)	Predominant frequency (PF)	Spectral acceleration for PF (g)	Amplification factor
BH-1	0.22	2.98	0.89	1.9
BH-2	0.19	3.61	0.67	1.48
BH-3	0.45	6.59	2.39	3.52
BH-4	0.34	9.33	2.21	2.64
BH-5	0.33	9.27	2.07	2.55
BH-6	0.34	3.71	1.40	2.64
BH-7	0.15	4.05	0.54	1.19
BH-8	0.25	5.03	0.75	1.79

variation of surface level PHA values obtained from SHAKE 2000 is presented in Fig. 5.13. The maximum PHA at the ground surface of the study area is 0.45 g, close to the location BH-3. The corrected SPT-N value for this location was found to be less than 20 at 3 m depth from the bore log data. As the depth of groundwater table is very low such a high value of the PHA aggravates liquefaction hazard in the location. Figure 5.14 presents the spatial variation of the amplification factor for PHA (corresponding to zero period) over the entire the Kalpakkam NPP site, where maximum amplification occurred at the location close to BH-3. Figures 5.15 and 5.16 present the amplification spectra for location BH-1 and BH-6 in the study area.

Predominant frequency is the frequency at which the maximum amplitude is observed for a ground motion in the frequency domain [181]. In this study the predominant frequency of the surface ground motion was also evaluated after converting it into frequency domain using Fast Fourier Transform (FFT) algorithm. The Fourier spectrum obtained from FFT fluctuates significantly and to avoid the influence of individual spikes, predominant frequency was obtained from a smoothened spectrum as shown in Fig. 5.17 [181, 360]. Hanning window [360]

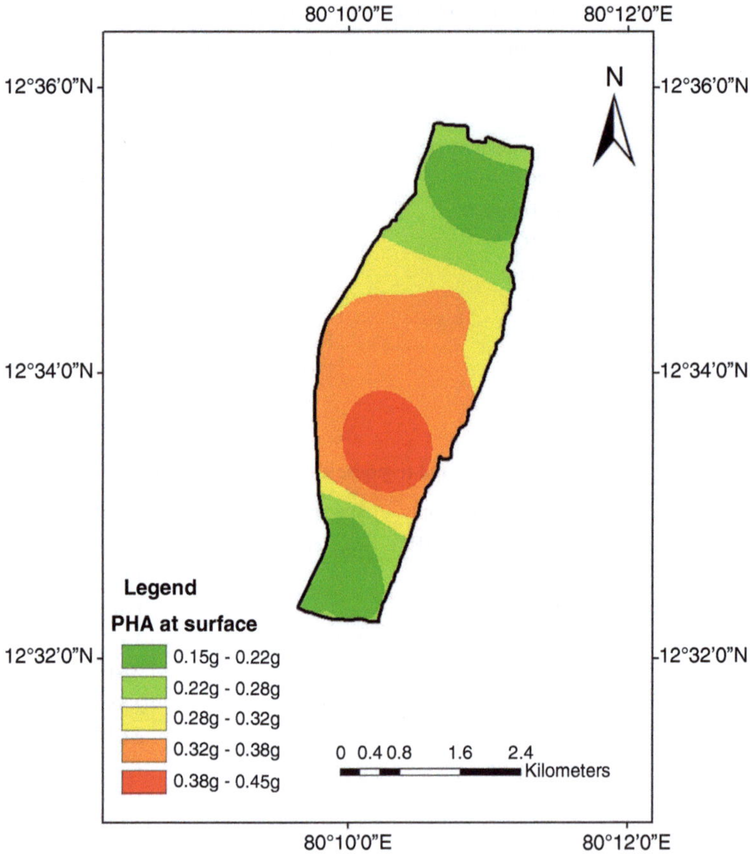

Fig. 5.13 Spatial variation of surface level PHA value throughout the Kalpakkam NPP site (after [153])

was adopted for smoothening the Fourier spectrum. Towhata [360] also suggested that significant amplification can be observed at places where the natural frequency of the soil is matched with the predominant frequency in the earthquake motion, leading to the extensive shaking and building damage.

The predominant frequency in the study area ranges from 2.9 to 9.3 Hz as given in Table 5.8 and its spatial variation is presented in Fig. 5.18. This chapter also presents response spectra of SDOF (single degree of freedom) for the amplified surface motion corresponding to 5% damping (Fig. 5.19) which is an important tool for a structural designer.

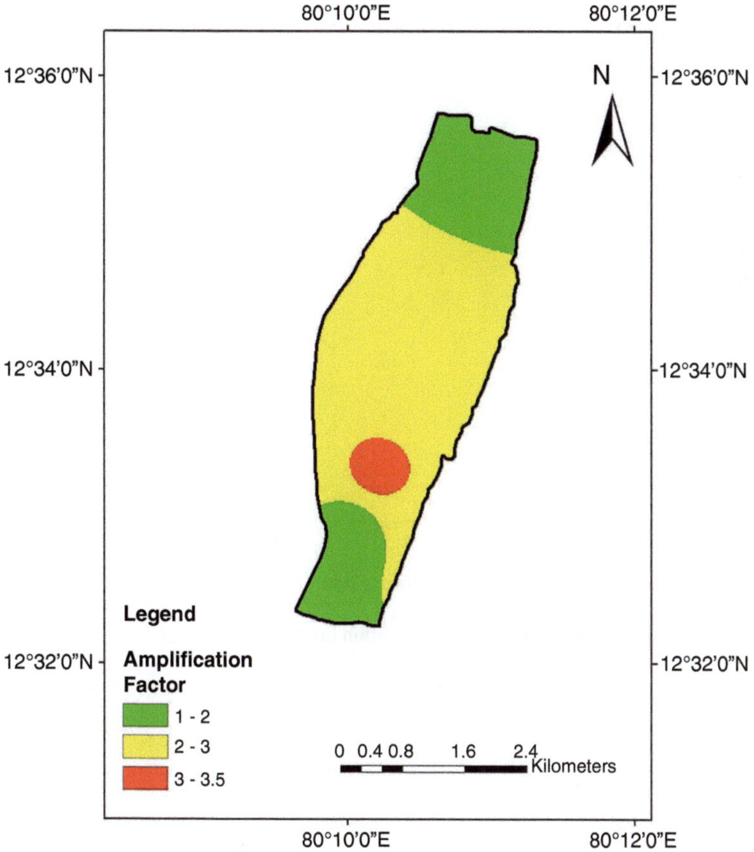

Fig. 5.14 Spatial variation of amplification factor throughout the Kalpakkam NPP site (after [153])

5.5 Evaluation of Site Amplification at Macro-Level: A Case Study of Karnataka State

For a small region, ground response analysis techniques are available for evaluating PHA at ground surface, however for a larger landmass such as a state or a country, this technique is not viable. Hence in the present study, for the state of Karnataka, the seismic hazard at ground surface was estimated using a non-linear site amplification technique proposed by Raghu Kanth and Iyengar [274]. Raghu Kanth and Iyengar [274] has developed a regression equation for estimating amplification factor for each NEHRP site class. Seismic site characterization for

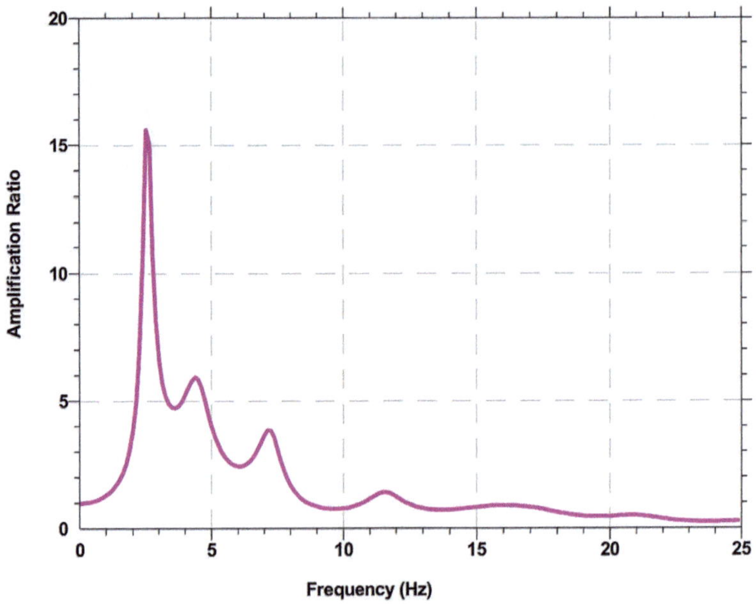

Fig. 5.15 Amplification spectrum for BH-1 location (after [153])

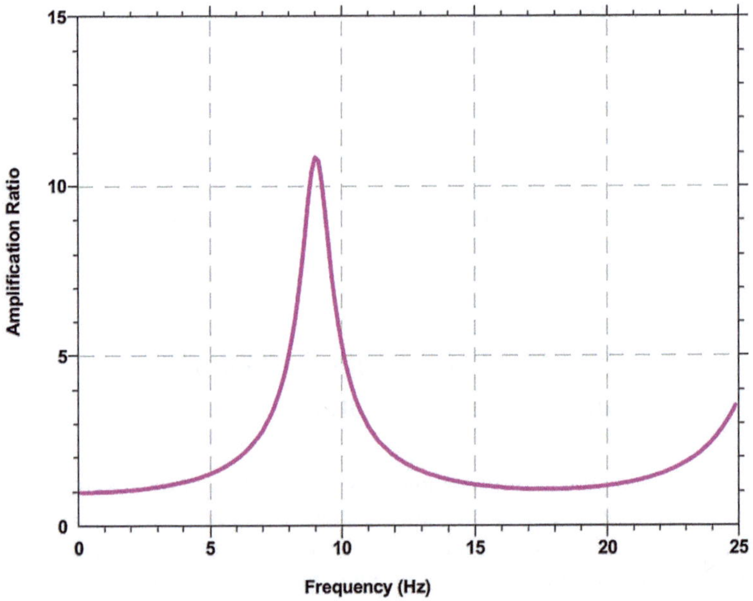

Fig. 5.16 Amplification spectrum for BH-6 location (after [153])

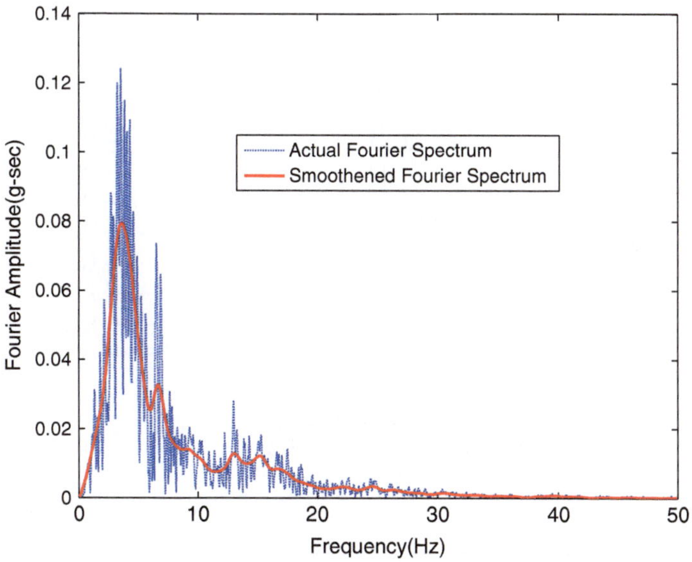

Fig. 5.17 Actual and smoothened Fourier spectrum of the surface ground motion (after [153])

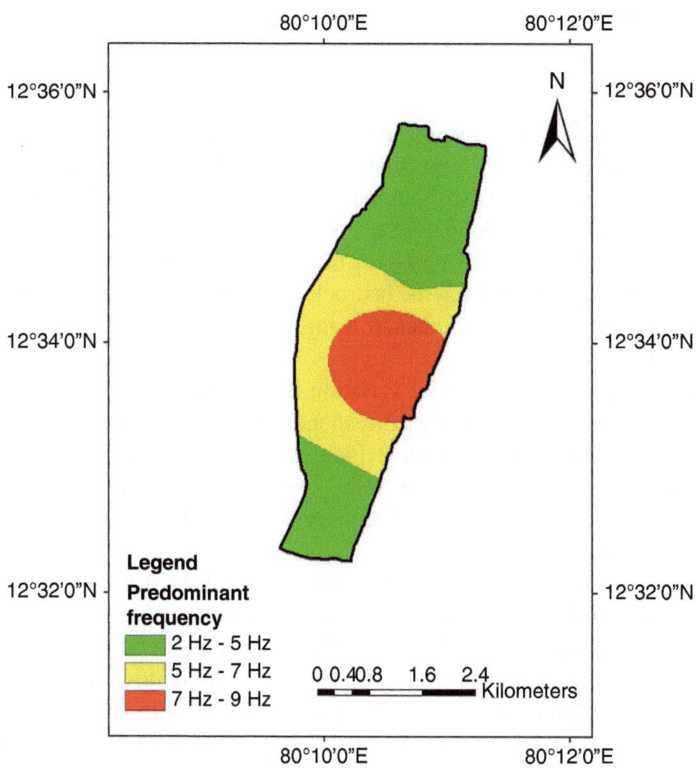

Fig. 5.18 Spatial variation of predominant frequency throughout the Kalpakkam NPP site (after [153])

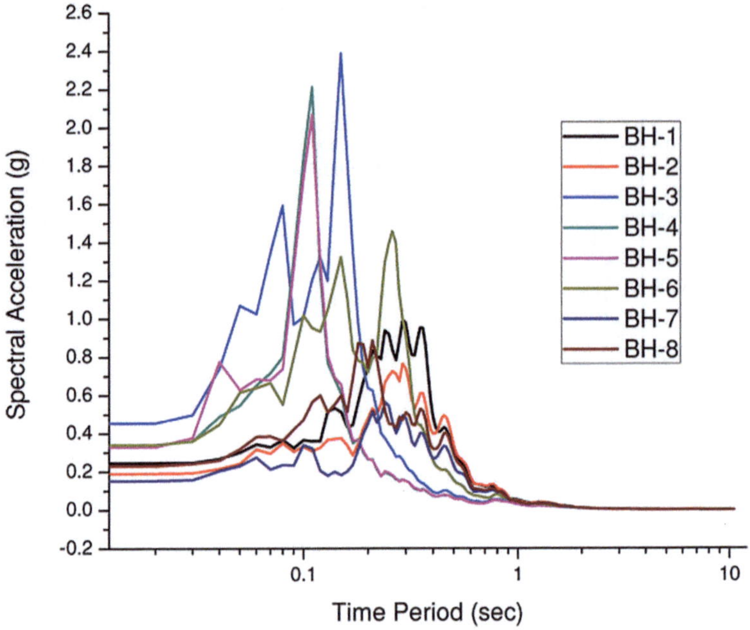

Fig. 5.19 Response spectrum plots at major locations in the study area (after [153])

the entire state of Karnataka was carried out using the satellite data, considering a grid size of 5 km × 5 km as mentioned in Sect. 4.4 of Chap. 4. The site amplification for each grid point was estimated by considering the NEHRP site class to which that grid point belongs, using the [274] methodology. Knowing the site amplification and bedrock PHA, the surface level PHA for each grid point was estimated by multiplying PHA at the bedrock and amplification factor. The bedrock level PHA at grid point was evaluated using both deterministic and probabilistic methodologies.

Figure 5.20 presents the spatial variation in the PHA value at ground surface throughout the state of Karnataka corresponding to bedrock level PHA obtained using the deterministic methodology. Figures 5.21 and 5.22 present the spatial variation in the surface level PHA value corresponding to bedrock level PHA obtained using the probabilistic methodology corresponding to 475 year and 2500 year return period.

Site amplification studies using bedrock PHA values from the deterministic seismic hazard analysis show the high surface level hazard value (0.4–0.55 g) for regions close to Bidar in north Karnataka. Similarly considering the bedrock level PHA values from probabilistic analysis, places in Bidar district have hazard value at surface ranging from 0.3 to 0.4 g, for a return period of 475 years and 0.4–0.52 g for a return period of 2500 years. Studies also show high surface level hazard in

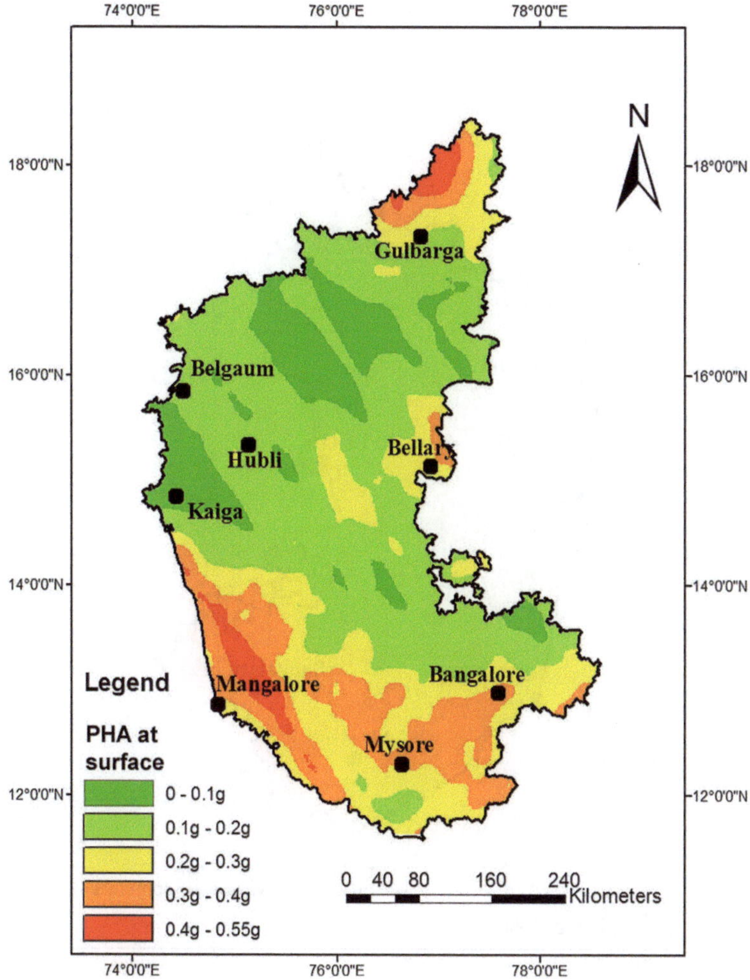

Fig. 5.20 Spatial variation of surface level PHA value throughout the Karnataka state from DSHA after [151])

the range of 0.4–0.55 g for the Mangalore–Udupi regions considering deterministic PHA values at bedrock. However considering the bedrock level PHA values from probabilistic method, the same region has moderately high surface level hazard ranging from 0.2 to 0.3 g for a return period of 475 years and 0.3–0.4 g for a return period of 2500 years.

The region between Bangalore and Mysore is also found to have moderately high value of surface PHA of about 0.3–0.4 g considering the bedrock level shaking from deterministic analysis. But, considering the bedrock level PHA values from

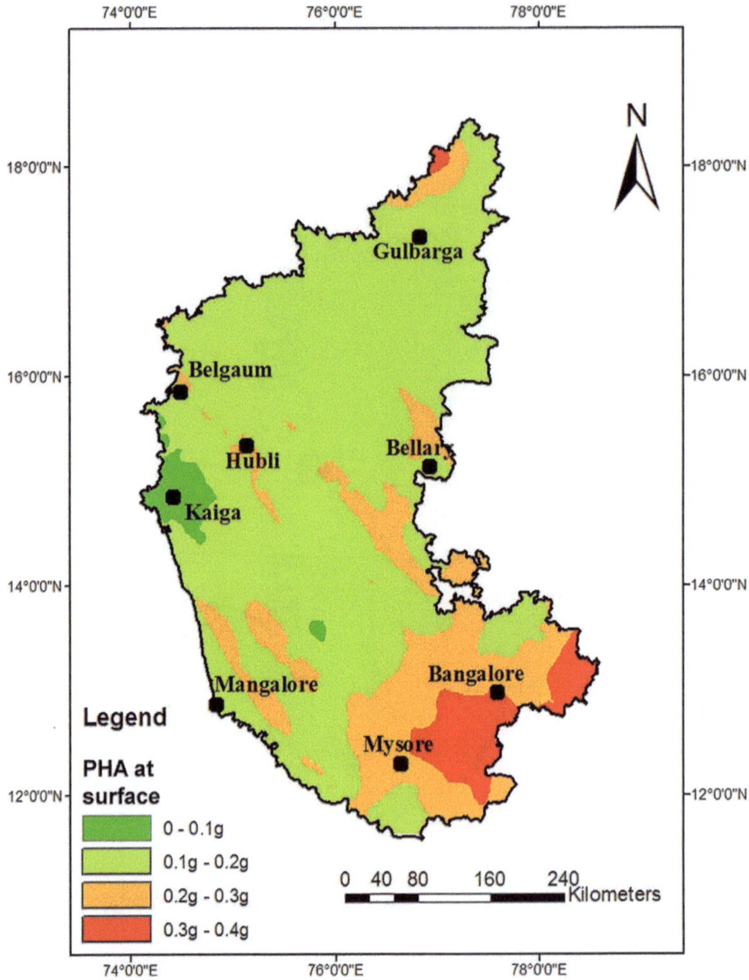

Fig. 5.21 Spatial variation of surface level PHA value throughout the Karnataka state from PSHA (for a return period of 475 year after [150])

probabilistic method, the same region is found to have high value of surface PHA of about 0.3–0.4 g for a return period of 475 year and 0.4–0.52 g for a return period of 2500 years. The interior parts of Karnataka state are found to have low surface level hazard value ranging from 0.01 to 0.2 g considering bedrock level PHA from deterministic method. The regions close to Bellary have moderate surface level hazard ranging from 0.2 to 0.3 g. Similar trends for PHA is also observed from

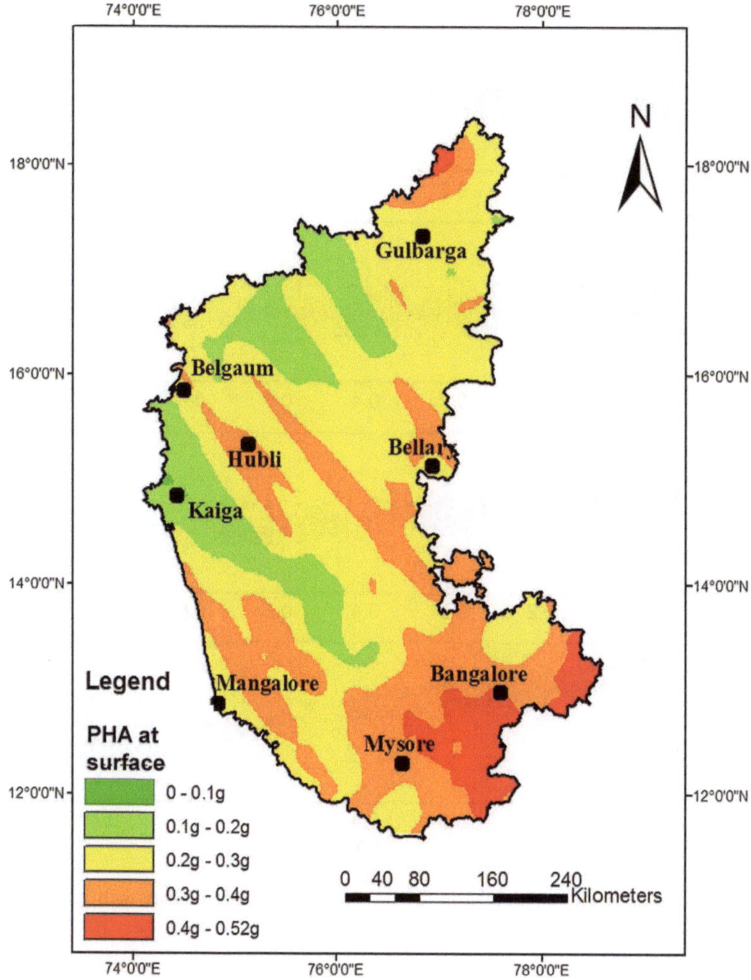

Fig. 5.22 Spatial variation of surface level PHA value throughout the Karnataka state from PSHA (for a return period of 2500 year after [151])

the site amplification studies considering the bedrock level PHA from probabilistic method having 475 year return period. However considering the bedrock level PHA having 2500 year return period, the surface level hazard value is in the category of moderately high to high.

Macro-level site amplification studies also show a low surface level hazard value for Kaiga, which is the location for nuclear power plant. The surface level PHA

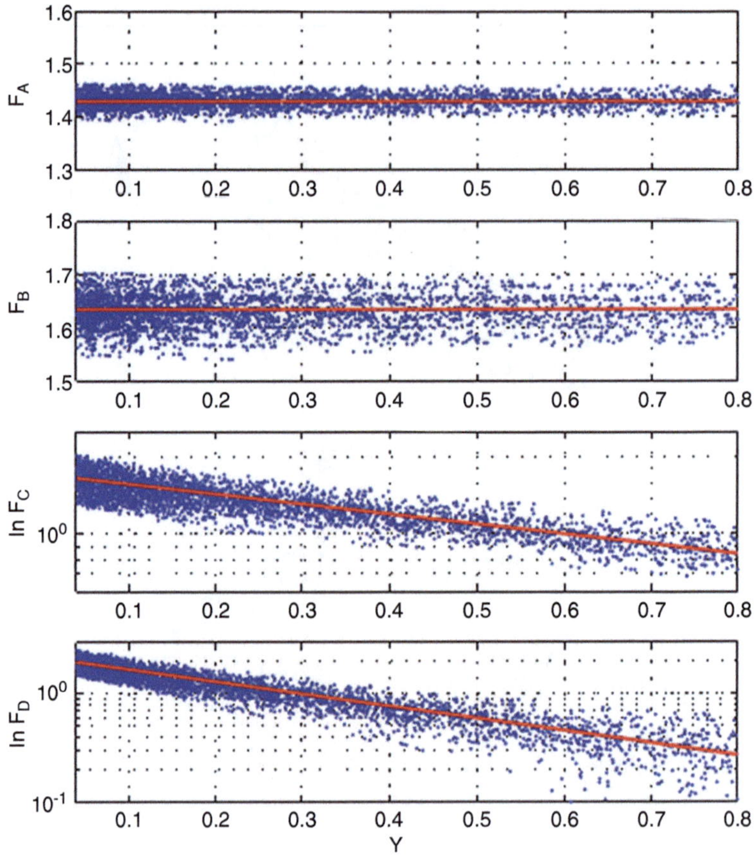

Fig. 5.23 Variation of site amplification factors for different site classes with rock level PHA values (after [274])

value for this region is 0.1 g. Considerable variations in the surface level PHA values for site class C and D is because the site coefficients (F_C and F_D) for C and D-type sites exhibit strong dependence on bedrock values when compared to (F_A and F_B) A and B type as shown in Fig. 5.23 given by Raghu Kanth and Iyengar [274].

Chapter 6
Liquefaction

6.1 Introduction

During an earthquake, soil can fail due to liquefaction with devastating effect such as land sliding, lateral spreading, or large ground settlement. The phenomenon of liquefaction of soil had been observed for many years, but was brought to the attention of researchers after Niigata (1964) and Alaska Earthquakes (1964). In India, classic scenario of earthquake induced soil liquefaction was observed during the Bhuj earthquake of 2001. Liquefaction is a phenomenon in which the strength and stiffness of a soil is reduced by earthquake shaking or by other dynamic loading. Liquefaction and related phenomena have been responsible for tremendous amounts of damage in historical earthquakes around the world [224, 391]. During the Bhuj earthquake on 26th January 2001 (M_W—7.7) lot of damages had been occurred due to liquefaction and other ground failures [276].

A large number of investigations have been carried out for understanding the phenomenon of soil liquefaction in the last four decades. From these investigations it was observed that a vast majority of liquefaction occurrences were associated with sandy soils and silty sands of low plasticity.

6.2 Mechanism of Soil Liquefaction

It is necessary to understand the mechanism of soil liquefaction, where it occurs and why it occurs so often during earthquakes. Figure 6.1 clearly depicts the mechanism of soil liquefaction. Liquefaction of soil is a process by which sediments below water table temporarily lose shear strength and behaves more as a viscous liquid than as a solid. The water in the soil voids exerts pressure upon the soil particles. If the pressure is low enough, the soil stays stable. However, once the water pressure exceeds a certain level, it decreases the effective stress and eventually the shear

T. G. Sitharam et al., *Comprehensive Seismic Zonation Schemes for Regions at Different Scales*, https://doi.org/10.1007/978-3-319-89659-5_6

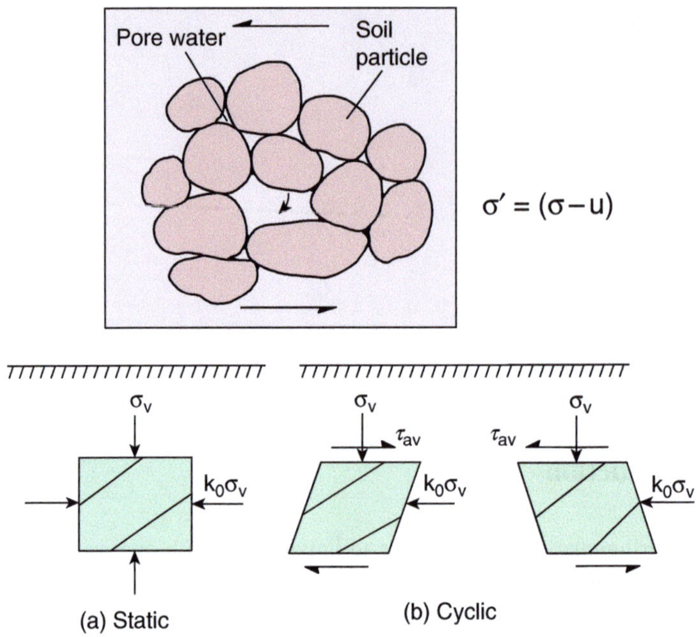

Fig. 6.1 Mechanism of soil liquefaction (after [238])

strength of the soil is reduced. During an earthquake, the sudden load is taken by the water in the form of excess pore water pressure. If increased pore water pressure cannot be released, it will continue to build up and it may lead to reduction of effective stress to zero. In this state, the soil layer loses its shear strength and it behaves like a viscous liquid and it cannot take the load coming from overlying soil or structures. Thus, the upper layers of soils move down and this will lead to the collapse of structures on the ground. Casagrande [57] proposed the concept of critical void ratio (CVR) line which is the locus of critical void ratios achieved by soils subjected to large strain shearing under drained, strain-controlled condition with different confining pressures. Saturated soils with initial states above CVR line (loose state) show tendency for contraction under drained loading hence produce positive excess pore pressure under undrained loading, and soils with initial states plotting below CVR line (dense state) show tendency for dilation under drained loading hence produce negative excess pore pressure under undrained loading. Thus CVR line was considered to mark the boundary between the soil states which are susceptible to flow liquefaction (loose), and soils which are nonsusceptible to flow liquefaction (dense). But, from investigations of partial failure of Fort Peck dam (September 1938) in Montana, it was found that sand located below the CVR line can also liquefy and it showed the inability of the type of test (strain controlled test) used for determining the critical void ratio to define correctly the boundary between sands that are safe and unsafe against flow liquefaction.

Later, Casagrande hypothesized the structure of flowing sand as that different from structure of sand in static state. In liquefied sand, each grain is constantly rotating in relation to all surrounding grains so as to offer a minimum of frictional resistance [58]. He termed this the flow structure. Castro [59] developed the flow structure of sand by conducting stress controlled undrained static and cyclic triaxial tests. He modified the critical void ratio line, termed as e_F line, as the locus of void ratios in which liquefaction with a flow structure developed which was plotted below and roughly parallel to the CVR line obtained from strain controlled tests. This difference in strength during the flow in strain controlled and stress controlled tests was attributed to the difference in strain rate. In stress controlled test, the constant load produces large strain rate during the flow and a perfect flow structure is developed, while in strain controlled test, the relatively low strain rate causes local groups of sand grains to lose temporarily their flow structure. Castro [59] observed limited liquefaction failure in medium dense sand. The specimen experienced a sudden failure with only a slight decrease in resistance. Then it exhibited a dilative behavior from this transformation point [143]. The test results obtained by Castro [59], showing liquefaction behavior (a), dilation behavior (c), and limited liquefaction (b) are illustrated in Fig. 6.2. Poulos [267, 268] termed the CVR line modified by Castro as steady state line (SSL). He defined the steady state of deformation as the state in which the soil flows continuously at constant void ratio, constant effective minor principal stress, and constant shear stress. The location of soil state with respect to SSL decides the susceptibility of soil to liquefaction. Soil whose state can be plotted above SSL is not susceptible to flow liquefaction and soil whose state is plotted below SSL is susceptible to flow liquefaction only if the static shear stress exceeds its steady state strength.

Seed and Lee [311] discovered that when a triaxial specimen of saturated sand is subjected to cyclic loading in an undrained test in such a manner that the principal stresses pass in each cycle through a hydrostatic state of stress (which means that

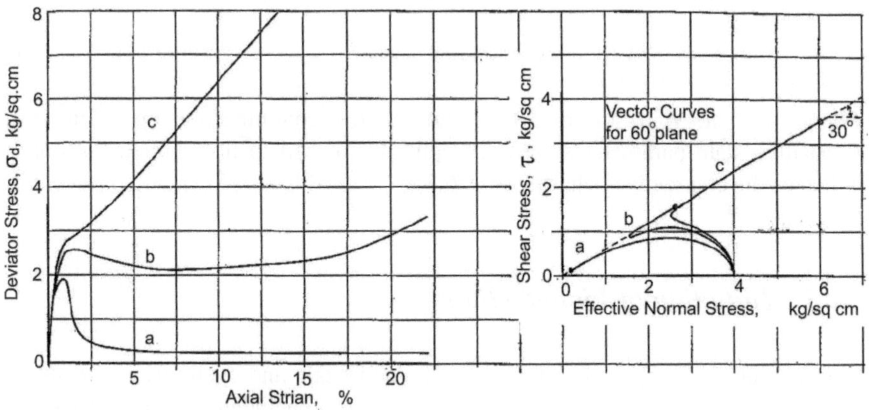

Fig. 6.2 Typical stress-strain curves from stress-controlled undrained static triaxial test (after [59])

Fig. 6.3 Definition of state parameter Ψ (after [22])

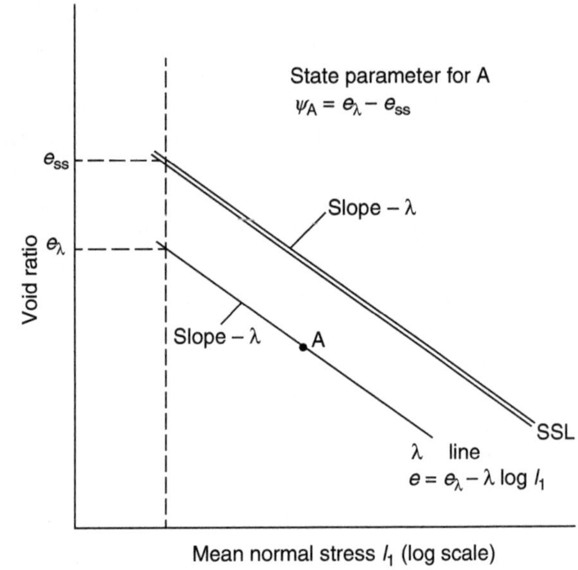

all shear stresses disappear), even a dense and highly dilative sand will develop cyclically high pore pressures and deformations. The stage at which pore pressure equals the confining pressure for the first time was called the initial liquefaction. Casagrande [58] called this type of liquefaction failure under cyclic loading as cyclic liquefaction or cyclic mobility.

From the concepts of critical-state soil mechanics, [22] coined the term state parameter that combines the influence of void ratio and stress level with reference to the ultimate (steady) state to describe sand behavior. The behavior of a cohesionless soil is more closely related to the proximity of its initial state to the critical-state line (steady state line). They defined the state parameter as per Eq. (6.1)

$$\Psi = e - e_{ss} \tag{6.1}$$

where e_{ss} is the void ratio of SSL at the effective confining pressure of interest. A positive state parameter shows susceptibility of soil to flow liquefaction as it indicates the contractive behavior (Fig. 6.3).

6.2.1 *Initiation of Liquefaction*

Hanzawa et al. [118], by conducting undrained static and cyclic triaxial tests on isotropically and anisotropically consolidated soils, concluded that the stress states corresponding to the peak strengths (stress states at which the samples suddenly collapse) fall on a straight line in stress path space passing through the origin, which

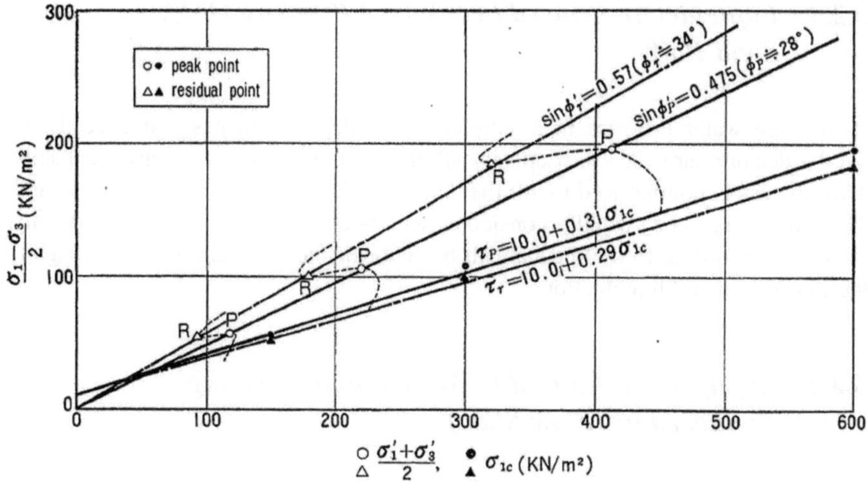

Fig. 6.4 FLS and SSL obtained from undrained static triaxial test conducted on anisotropically consolidated soil (after [118])

is termed as flow liquefaction surface (FLS) and is illustrated in Fig. 6.4. In Fig. 6.4, the peak points show the flow liquefaction line and residual points represents the steady state of deformation. The same results were produced by Vaid and Chern [370, 371]. Once the soil collapses it develops a flow structure and fall to steady state of deformation possessing its residual strength. The closer the initial stress state of soil to the FLS, the susceptible is the soil to flow liquefaction. An analogous conceptual surface was proposed by [337] and was named as collapse surface which passes through the steady-state point. Yang [392] experimentally proved that the flow liquefaction line is not a material constant but rather varies with the state of soil. In the framework of critical-state soil mechanics, he suggested an explicit relationship between the slope of FSL and the state parameter that accounts for both stress level and the density of soil.

Sitharam and Dinesh [330] and Dinesh et al. [77] numerically generated qualitatively realistic macroscopic cyclic behavior of granular materials using discrete element model (DEM) introduced by [74]. The DEM analysis exhibited an increase in resistance to liquefaction with increasing confining pressure since the separation between the individual particles is prevented. It was also observed that loose samples liquefy during cyclic loading under undrained conditions at much smaller cyclic deviator stress level than the deviator stress they carry under monotonic shearing. El Shamy and Abdelhamid [84] conducted similar numerical studies by idealizing solid phase using discrete element method and fluid phase using lattice Boltzmann method. They suggested that liquefaction is due to the reduction in void space during shaking that leads to build up in pore-fluid pressure.

6.2.2 Liquefaction Caused by Seepage Pressure Only: Sand Boils

If the pore water pressure in a saturated sand deposit reaches and exceeds the overburden pressure, the sand deposits will float or boil and lose its bearing capacity. This process is nothing to do with the density and volumetric contraction of sand. Therefore, it has been usually considered as a phenomenon of seepage instability. However, according to the mechanism behavior of the material, it also belongs to the category of soil liquefaction.

6.2.3 Liquefaction Caused by Monotonous Loading or Shearing: (Flow Slide)

The concept of critical void ratio has been suggested by Casagrande. The skeleton of loose saturated sand exhibits irreversible contraction in bulk volume under the action of monotonous loading or shearing, which will cause increase of pore water pressure and decrease of effective stress and finally brings about an unlimited flow deformation.

6.2.4 Liquefaction Caused by Cyclic Loading or Shearing: Cyclic Mobility

With various experimental techniques and testing apparatus it has been found out that cohesionless soil always show volumetric contraction at low shear strain level, but may dilate at higher shear strain level depending upon the relative density of soil. Therefore, under the action of cyclic shearing a saturated cohesionless soil could show liquefaction at time intervals when shear strain is low, but may regain shear resistance in time intervals when the shear strain level is higher. A sequence of such sort of intermittent liquefaction would bring about the phenomenon of cyclic mobility with limited flow deformation. If the saturated cohesionless soil was loose enough to keep contraction at high shear strain level, then it also could come out to be an unlimited flow deformation.

6.3 Liquefaction Susceptibility

There must have been certain conditions in order to soil to liquefy during an earthquake. Hence it is the task of geotechnical engineers to determine whether the given soil strata is susceptible to liquefaction during any earthquake. The liquefaction susceptibility of any soil strata is determined by considering the in-situ soil properties only. Following are the important factors that govern the liquefaction susceptibility of any soil.

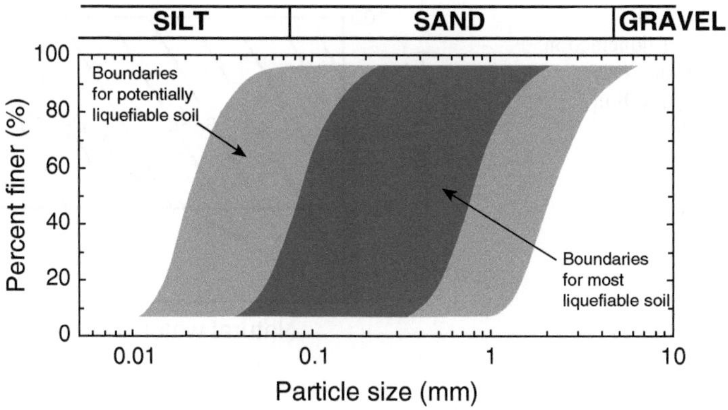

Fig. 6.5 Grain size boundary for defining most liquefiable and potentially liquefiable soils (after [363])

6.3.1 *Type of Soil (Index Properties of Soil)*

The phenomenon of liquefaction is most commonly observed in loose, cohesionless soils. Earlier it was believed that liquefaction was limited to sands only. As the grain size becomes coarser the liquefaction resistance increases due to increase in the weight of soil particles and improved drainage. However, now it has been found out that liquefaction takes place in gravelly soils as well [398]. Consequently, clarifying the gradation curve of liquefiable soil is an important approach to liquefaction susceptibility of a ground. Tsuchida [363] already showed the ranges of grain size distribution of liquefiable soil in 1970. These ranges are used in the Technical standards for Port and Harbour Facilities published by the Japan Port and Harbour Association (Fig. 6.5). Further, it was considered that fine grained soils were incapable of generating the high pore pressure. Lab experiments have shown that majority of clays remain unsusceptible to liquefaction. Sensitive clays can exhibit strain softening behavior which makes it vulnerable to liquefaction. Ishihara [138] suggested that liquefaction susceptibility of fine grained soils like silts and clays is mainly influenced by liquid limit and water content which was presented in Fig. 6.6 as per [209]. If the soil samples do not fall in the safe zone (as in Fig. 6.6) based on its natural moisture content and liquid limit, the liquefaction susceptibility of such soil sample must be verified using undrained cyclic loading tests.

6.3.2 *Shape of Soil Particles*

Rounded particles are known to density more easily than angular shaped particles. Hence, the ease of densification is directly proportional to the susceptibility of soil to liquefaction. This implies that rounded particles are more prone to liquefaction than angular shaped particles, during an earthquake.

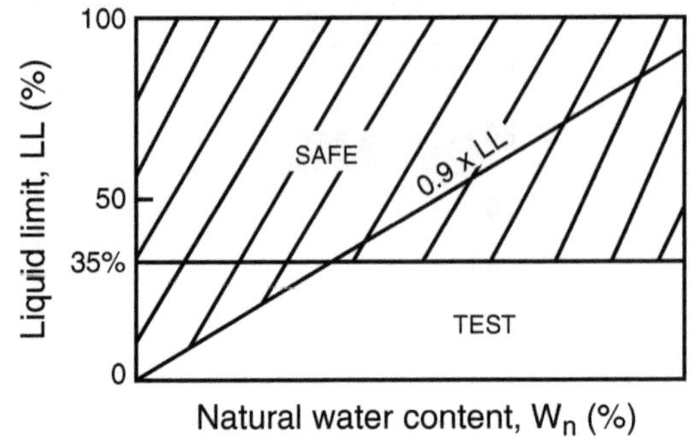

Fig. 6.6 Graphical representation of liquefaction criteria for silts and clays from studies by [380] in China (after [209])

6.3.3 *Permeability of Soil*

Permeability of soil mass leads to dissipation of pore water pressure very quickly. Hence, if this dissipation of pore water pressure occurs very quickly, then the soil may not liquefy. Based on the observations, it can be concluded that presence of highly permeable soil layers can reduce liquefaction potential of adjacent soil layers. In other words, the liquefaction susceptibility of a loose sand layer will be reduced if it is surrounded by a highly permeable gravel layer.

6.3.4 *Presence of Seismic Waves*

An earthquake is characteristically a kind of vibration energy that includes impacts or shocks. The vibration it produces varies according to the characteristics of the ground through which the vibrations transmit. Earthquake on the surface of the ground is strongly dependent on the vibration characteristics of the layers underneath.

It is known that the most common cause for liquefaction is the presence of vibration/seismic energy released during an earthquake. Based on the observations, it has been concluded that the potential for liquefaction increases with the increase in seismic energy. The energy thus produced is the cause for the cyclic shear stress that acts on the soil mass and this causes the acceleration of the soil mass in the horizontal direction. The acceleration is known as the peak acceleration exerted by earthquake at the ground surface. It has been stated that the shaking threshold required for liquefaction is a local shaking magnitude of about five and a peak acceleration of 0.1 g.

6.3.5 *Depth of Ground water Table*

Another criterion which makes soil susceptible to liquefaction is its degree of saturation. This clearly indicates the importance of depth of water table and it must be near the surface of the ground. The soil located above the water table is unsaturated and hence it will not liquefy. For those locations where the depth of water table is very deep, the liquefaction susceptibility will be less. Moreover if the level of the groundwater table keeps on changing the liquefaction susceptibility of the soil will also fluctuate.

6.3.6 Historical Environment

Observations from earlier earthquakes in the region can offer a great amount of information regarding the liquefaction susceptibility of the soil mass at the site. If the soil has already liquefied in the past due to an earthquake, then it can liquefy again in the future earthquakes. The information on the past liquefaction can be obtained in the form of maps, which provides information on the previous earthquakes (and the earthquakes expected to occur in the future at the site) which caused the liquefaction of soil. It has also been observed that liquefaction effects are historically confined to a zone that lies within a particular distance from the tremor source. It is also obvious that the distance to which the effect of liquefaction can be expected is directly related to the magnitude of the earthquake.

6.3.7 Age of Soil

The age of the soil plays a pivotal role in determining the liquefaction susceptibility. The liquefaction resistance of a soil increases, with the age of the soil. New soils liquefy more easily than old soil deposits, because the old soil deposits have more stable particle arrangement due to the compression of the soil particles with time. Moreover the particle bonding in the case of old deposits will be more. Thus, old soil deposits have a greater liquefaction resistance comparatively.

6.3.8 Confining Pressure

Soils at depths of more than 15 m generally do not liquefy due to the high confining pressure on it. The confining pressure is the overburden pressure acting at a point in the soil mass due to the surcharge by the soil layers above that point. More the confining pressure, lesser is the probability of the soil to liquefy. As already mentioned, if the groundwater table depth is more than 15 m from the ground surface, the soil lying below the water table generally does not liquefy. But this doesn't stop us from determining the liquefaction susceptibility of the soil layer that is at a depth of more than 15 m from the ground surface. In case of a sloping dam, the soil deposit has to be analyzed for its liquefaction susceptibility for a thickness which lies in water, even if this thickness lies at a depth of more than 15 m.

6.3.9 Relative Density

The soil layer which has a low relative density is more susceptible to liquefaction than dense soil. Loose soils density easily during earthquake and this will cause an increase in pore water pressure which leads to the liquefaction of soil, where as dense sands will dilate during earthquake and this will reduce the pore water pressure and hence the liquefaction susceptibility will be less.

6.3.10 Natural Soil Deposits in Water Bodies

Soils in lakes, rivers, or oceans are highly prone to liquefaction due to their loose and segregated structure. Hence soils which are susceptible to liquefaction are formed in marine depositional environments.

6.4 Liquefaction Susceptibility Analysis Based on SPT and CPT

Major in-situ tests such as the standard penetration test (SPT), cone penetration test (CPT) are most useful to determine the liquefaction susceptibility of the soil. Based on the studies of [285, 288], a soil classification based on CPT for assessing liquefaction susceptibility of fine grained soil as in Fig. 6.7. The quantities normalized CPT penetration resistance Q (dimensionless) versus normalized friction ratio F was generated first, where Q and F are defined as per Eqs. (6.2) and (6.3).

$$Q = \frac{q_c - \sigma_{vo}}{\sigma'_{vo}} \tag{6.2}$$

$$Q = \frac{f_s}{q_c - \sigma_{vo}} \tag{6.3}$$

where q_c and f_s are the tip resistance and sleeve friction from the cone penetration test. The zone A in Fig. 6.7 represents the region where cyclic liquefaction is possible depending on size and duration of cyclic loading. Soil samples whose tip resistance and sleeve friction combination lies in zone B is unlikely to liquefy, check other criteria; zone C, flow liquefaction and (or) cyclic liquefaction possible, depending on soil plasticity and sensitivity as well as size and duration of cyclic loading

Fig. 6.7 Soil classification chart by Robertson and Wride [288]

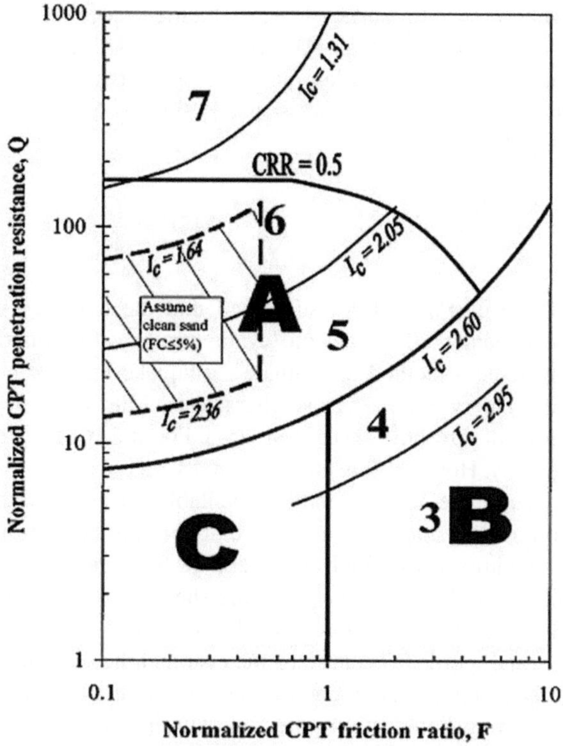

6.5 Evaluation of Liquefaction Potential

The liquefaction potential of soil significantly depends on the initial state of the soil as well as on the intensity of earthquake loading. The earthquake loading is the driving force causing liquefaction, while the soil resistance against liquefaction mainly depends upon the initial state of the soil. Hence the evaluation procedure for liquefaction potential requires two steps,

- Characterizing earthquake loading
- Characterizing soil resistance

Seed and Idriss [309] has proposed a simplified approach for characterizing the earthquake loading which is most popular. In this approach, the earthquake loading is characterized in terms of cyclic shear stress (or cyclic stress ratio) as presented in Eq. (6.4).

$$\text{CSR} = 0.65\left(\frac{a_{\max}}{g}\right)\left(\frac{\sigma_{v0}}{\sigma'_{v0}}\right)\left(\frac{r_d}{\text{MSF}}\right) \tag{6.4}$$

where $a_{\max}$ is the surface level peak ground acceleration (PHA) at the given borehole location obtained using deterministic hazard analysis; σ_{v0} and σ'_{v0} are the total and effective overburden pressure; r_d—depth reduction factor used to account for the flexibility of the soil as per [193] and MSF is the magnitude scaling factor (Eq. (6.5)),

$$\mathrm{MSF} = \frac{10^{2.44}}{M_W^{2.56}} \tag{6.5}$$

However [78] have suggested that the tendency of pore pressure generation is fundamentally related to cyclic strains rather than cyclic stresses. Hence they proposed a method to evaluate earthquake loading in terms of cyclic shear strain (γ_c). Dobry et al. [78] also characterized the soil resistance against liquefaction in terms of threshold shear strain (γ_t). Possibility of liquefaction will be maximum if $\gamma_c > \gamma_t$. However this method is not very popular.

There are several methods available for evaluating the cyclic resistance of the soil against liquefaction. Important ones are laboratory tests and in-situ field tests. Laboratory tests which are available for measuring soil resistance are cyclic triaxial tests [59, 93], cyclic simple shear test [272], and cyclic torsional test [142]. However the evaluation of cyclic resistance using laboratory tests is restricted to research purpose because of the cost and the difficulties involved in getting undisturbed samples. As there are many difficulties and the associated laboratory tests, evaluating liquefaction potential using in-situ tests has become more convenient. Major geotechnical field tests used for evaluating the cyclic resistance of soil are (1) Standard Penetration Test [131, 313, 315], (2) Cone penetration tests [286, 306], and (3) Based on shear wave velocity [346, 355].

6.5.1 Liquefaction Potential Based on Lab Tests

Cyclic Triaxial Test

The cyclic triaxial test is the most commonly used test for the measurement of dynamic soil properties at high strain levels. This test simulates the liquefaction phenomenon during earthquakes by applying cyclic shear to the saturated sandy soil under undrained condition. Axial loading is applied in steps to the specimen and the shear strain and shear stress at various intervals of loading are measured. Five to ten levels of shear strain amplitudes are chosen from the range of 10^{-5} to 10^{-2} for testing. Dynamic deformation characteristics are influenced by the effective confining pressure during the test. When an undisturbed sample of normally consolidated soil is obtained, the effective vertical pressure at the depth of sampling is isotropically applied by cell pressure to avoid the influence of over consolidation. In order to obtain in-situ shear modulus of the soil from the laboratory test results, correction of these results is necessary so that a shear modulus corresponding to the average effective principal stress at the sample depth is obtained. Seed and Lee [311] were the first to reproduce liquefaction in a cyclic triaxial test on loose and dense sands and concluded that liquefaction occurs more easily in sandy soils having higher void ratios. In addition to this they have proved that if the effective confining pressure is less than the liquefaction susceptibility will be high.

Cyclic Simple Shear Test

The cyclic direct simple shear test is capable of reproducing earthquake stress conditions much more accurately than the cyclic triaxial test. It is most commonly used for liquefaction testing. In this test, a short cylindrical specimen is restrained against lateral expansion by rigid boundary platens, a wire reinforced membrane or with a series of stacked rings. By applying cyclic horizontal shear stresses to the top or bottom of the specimen, the test sample is deformed in the same way as an element of soil subjected to vertically propagating S waves. In recent years, simple shear devices that allow independent control of vertical and horizontal stresses have been developed. To simulate the actual earthquake conditions, [272] used a large-scale simple shear apparatus. It was found that cyclic strength is related to the relative density of the soil and cyclic stresses that cause liquefaction in simple shear were less than those causing liquefaction in triaxial shear.

Cyclic Torsional Shear Test

Many of the difficulties with cyclic triaxial and cyclic shear test can be overcome with cyclic torsional shear test. This is mostly used to determine stiffness and damping characteristics over a wide range of strain levels. It allows isotropic or an isotropic initial stress conditions and can impose cyclic shear stresses on horizontal planes with continuous rotation of principal axes. Dobry et al. [78] used strain controlled cyclic torsional loading along with stress controlled axial loading of solid specimens and has proven effective for measurement of liquefaction behavior. Torsional testing of solid specimens, however, produces shear stresses that range from zero along the axis of the specimen to a maximum value at the outer edge. To increase the radial uniformity of shear strains, a hollow cylindrical cyclic torsional shear apparatus was also developed. Hollow cylinder tests offer perhaps the best uniformity and control over stresses and drainage. Ishihara and Li [142] developed a torsional triaxial shear test and conducted strain controlled tests on solid cylinders of saturated sands. These tests helped in establishing relationship between cyclic triaxial tests, cyclic simple shear tests, and the torsional triaxial test.

Shake Table Test

Shake table tests of many sizes are being used for liquefaction studies on saturated soil samples. The sample is prepared in a container, fixed to a shaking platform and vibrated at the desired frequency for a prescribed time. A surcharge is placed on the sample to provide the confining pressure. The measurements of acceleration, pore water pressure, and settlements are made during the test.

6.5.2 Liquefaction Potential Based on Field Tests

The difficulties and the high cost involved in the laboratory test make the in-situ-tests convenient method for modelling liquefaction potential. The four major in-situ test methods which are considered for the liquefaction potential evaluation are given below.

1. The Standard Penetration Test (SPT)
2. The cone penetration test (CPT)
3. Measurement of in-situ shear wave velocity (Vs)
4. The Becker penetration test (BPT).

SPT Based Methods

Standard penetration test is widely used as an economical, quick, and convenient method for investigating the penetration resistance of non-cohesive soils. The use of SPT as a tool for evaluation of liquefaction potential began to evolve in the wake of a pair of devastating earthquakes that occurred in 1964; the 1964 Great Alaskan Earthquake ($M = 9.2$) and 1964 Niigata Earthquake ($M = 7.5$), both of which produced significant liquefaction related damage. The SPT values obtained from the site investigation must be corrected to remove the errors. The N-values measured in the field were subjected to various corrections such as (1) Overburden pressure correction (C_N), (2) Hammer energy correction (C_E), (3) Borehole diameter correction (C_B), (4) liner correction (C_S), (5) Rod length correction (C_R) and (6) correction for fine contents (C_{FC}) [61, 313, 336, 400]. The SPT value corrected for 60% energy efficiency and 100 kPa overburden pressure $(N_1)_{60}$ was evaluated using Eq. (6.6) [287]. The $(N_1)_{60cs}$ provides consistent value for penetration resistance and hence become a standard for assessing liquefaction potential [291].

$$(N_1)_{60} = N \times [C_N \times C_E \times C_B \times C_S \times C_R] \tag{6.6}$$

The obtained $(N_1)_{60}$ was again corrected for fine content as suggested by Idriss and Boulanger [131] using Eqs. (6.7) and (6.8).

$$(N_1)_{60cs} = (N_1)_{60} + \Delta(N_1)_{60} \tag{6.7}$$

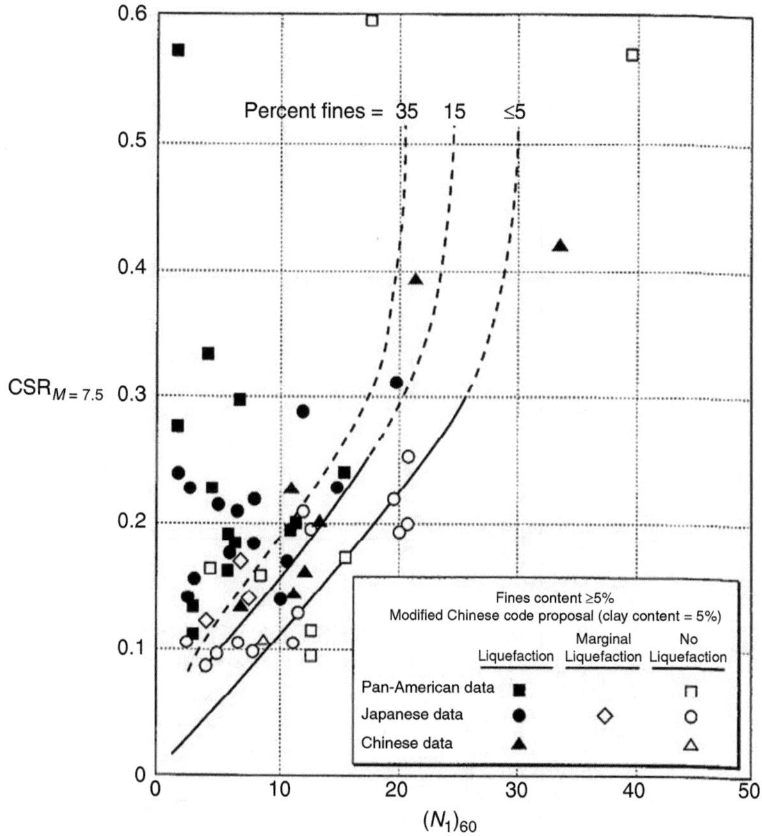

Fig. 6.8 Relationship between cyclic stress ratios causing liquefaction and $(N_1)_{60}$ values for clean sands for $M_7.5$ earthquakes (Seed et al. [305])

where

$$\Delta(N_1)_{60} = \exp\left[1.63 + \left(\frac{9.7}{FC+0.1}\right) - \left(\frac{15.7}{FC+0.1}\right)^2\right] \tag{6.8}$$

Here FC is the fine content, i.e. percentage of dry weight finer than 75 μm.

The most widely used methods to evaluate the liquefaction potential based on SPT values were proposed by Seed et al. [315]. This relationship is between corrected SPT-N values, corrected for overburden stress, instrument errors and other factors that affect the SPT testing, vs. intensity of cyclic loading that is expressed in the form of uniform cyclic stress ratio (CSR) (see Fig. 6.8). This was further modified by Youd et al. [400] and the curves between cyclic resistance ratio (CRR) and the corrected SPT values are given in Fig. 6.9. In this work it was noted that the CRR value will increase with fines content. Although this method is widely used, it is outdated and doesn't accommodate conditions by earthquakes that produce very

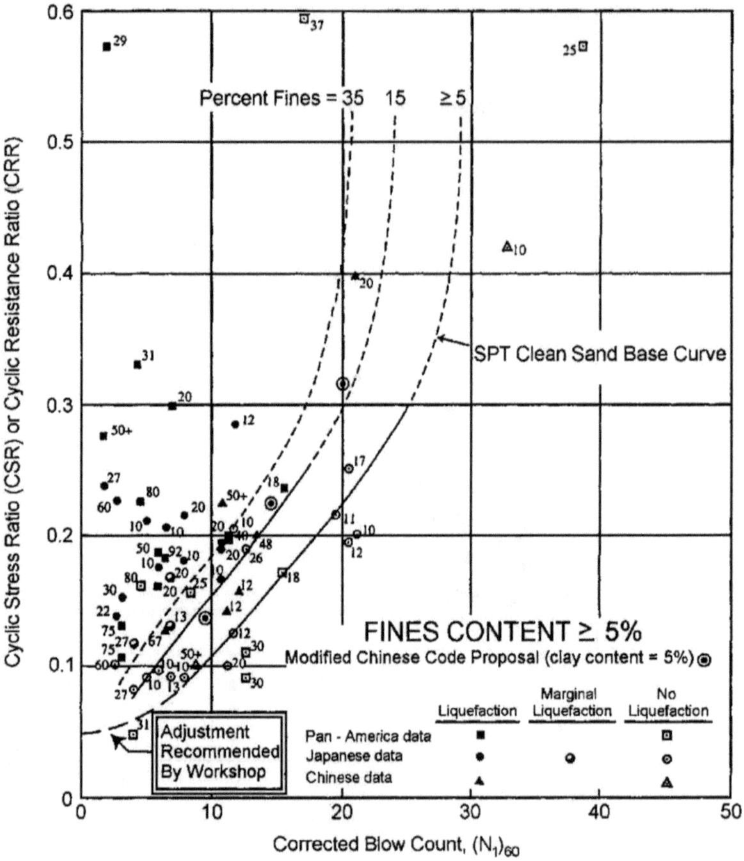

Fig. 6.9 Deterministic cyclic resistance curves proposed by Youd et al. [400]

high peak ground shaking levels (CSR > 2.5). The relationship does not have a formal probabilistic approach as well; i.e. the relationship does not provide the probability or uncertainty of occurrence of liquefaction.

A number of researchers have developed correlations based on probabilistic methods to evaluate the liquefaction potential. One of the first to develop a relation was [194] (Fig. 6.10) and more recently [399] (Fig. 6.11), and [358] (Fig. 6.12). The relationships provided by these researchers are in the form of probability contours (probability of triggering soil liquefaction). In the following figures these curves are superimposed over the [314] deterministic relationship for comparison.

The [399] correlation is based on the seismic data that were used by Seed et al. [314], which at that time were of very low quality. This relationship was developed by using the binary regression of logistic models method, for probabilistic regression. The likelihood function prepared, overstated the overall variance or uncertainty of the proposed correlation. An additional inadequacy was that [194]

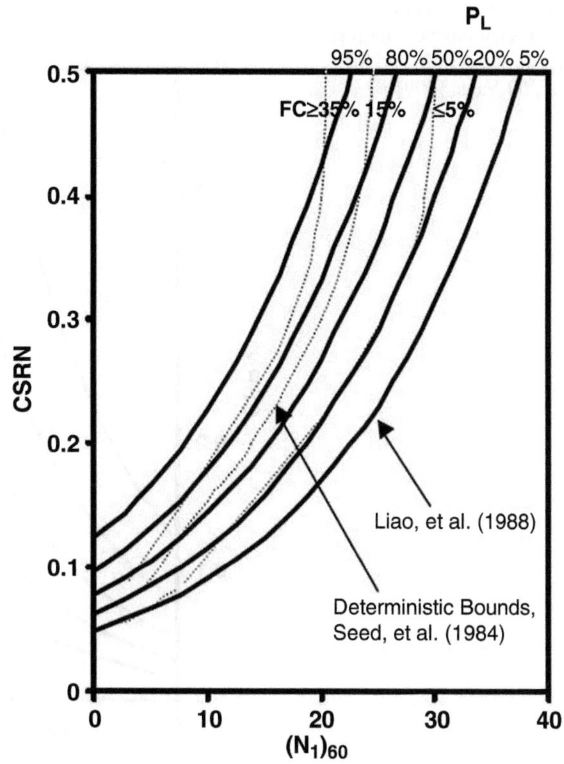

Fig. 6.10 Correlation between equivalent uniform cyclic stress ratio and SPT $(N_1)_{60}$ value [194]

couldn't find the effect of fines content in the soil on the regression relationship between SPT-N values and liquefaction resistance. So the relation developed was for sandy soils with less than 12% fines.

Youd et al. [399] correlation takes into account a number of field case history data points from the past earthquakes which have occurred since the earlier relationships have been developed. The correlation also excludes the limitations of the [194] relationship as discussed above. This correlation is applicable to soils of varied fine contents and hence can be used for silty soils as well.

The correlation as proposed [358] also makes use of an extended field case history database, and excludes the most of the questionable data used by Liao et al. [194]. The regression method used here is also that of binary regression, and as a result the overall uncertainty is again very large. However in spite of all these shortcomings, all the above relationships discussed here are being widely used.

Idriss and Boulanger [131] have revised SPT and CPT based semi-empirical relations for evaluating the liquefaction potential of saturated cohesionless soils. Further, they have also modified stress reduction factor (r_d), earthquake magnitude scaling factor for cyclic stress ratios (MSF) and overburden correction factor for cyclic stress ratios (K_σ), and the overburden normalization factor for penetration resistances (C_N).

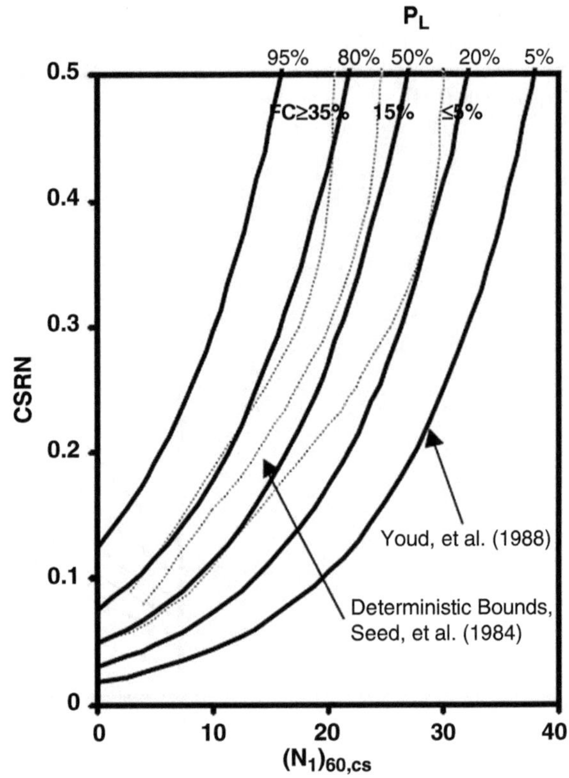

Fig. 6.11 Correlation between equivalent uniform cyclic stress ratio and SPT N1, 60-Value [399]

The cyclic resistance ratio (CRR) for each layer was evaluated using a methodology proposed by Idriss and Boulanger [131] given by Eq. (6.9)

$$\mathrm{CRR} = \exp\left[\left(\frac{(N_1)_{60cs}}{14.1}\right) + \left(\frac{(N_1)_{60cs}}{126}\right)^2 - \left(\frac{(N_1)_{60cs}}{23.6}\right)^3 + \left(\frac{(N_1)_{60cs}}{25.4}\right)^4 - 2.8\right] \tag{6.9}$$

where$(N_1)_{60cs}$ is the SPT-N value corrected for 60% energy efficiency, 100 kPa overburden pressure and fine content.

CPT Based Methods

In the previous section the discussion was centered around the correlations to evaluate the liquefaction potential based on SPT values. Due to the better accuracy and repeatability, several correlations are available for estimating the CRR values based on CPT values. The first to propose liquefaction triggering model based on CPT values was [286, 306]. One of the most widely used correlations nowadays is the one proposed by Robertson and Wride [288] (Fig. 6.13).

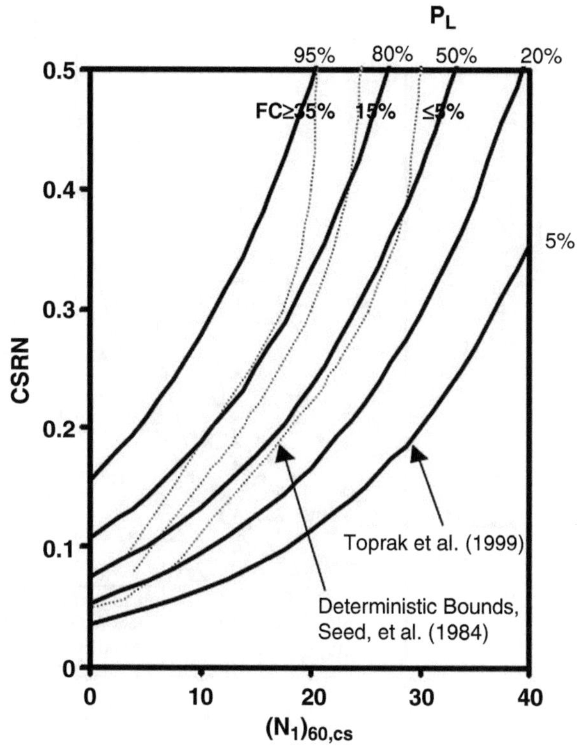

Fig. 6.12 Correlation between equivalent uniform cyclic stress ratio and SPT N1, 60 value [358]

6.5.3 Probabilistic Methodology

One of the major shortcomings in the deterministic method is that it considers only one critical magnitude-acceleration scenario for the evaluation of liquefaction potential. As it is evident that the liquefaction hazard at a site is not solely function of single ground acceleration and earthquake magnitude ($a_{\max}$ and M_W), but several magnitudes acceleration scenarios contribute towards the liquefaction hazard and their percentage of contribution varies. Hence, a performance based approach suggested by Kramer and Mayfield [182] was also adopted for the evaluation of liquefaction hazard at micro-level, where the contributions from all magnitudes and all acceleration levels were considered. Based on the available SPT data, the annual probability of exceedance for a given factor of safety value at a particular depth in a borehole (Λ_{FS^*}) is estimated using Eq. (6.10) [182].

$$\Lambda_{\mathrm{FS}^*} = \sum_{i=1}^{N_a} \sum_{j=1}^{N_m} P[\mathrm{FS} > \mathrm{FS}^* | a_i, m_j] \Delta\lambda_{a_i, m_j} \tag{6.10}$$

where N_a and N_m are the number of acceleration and magnitude increment; a_i and m_j are the incremental values in acceleration level and earthquake magnitude,

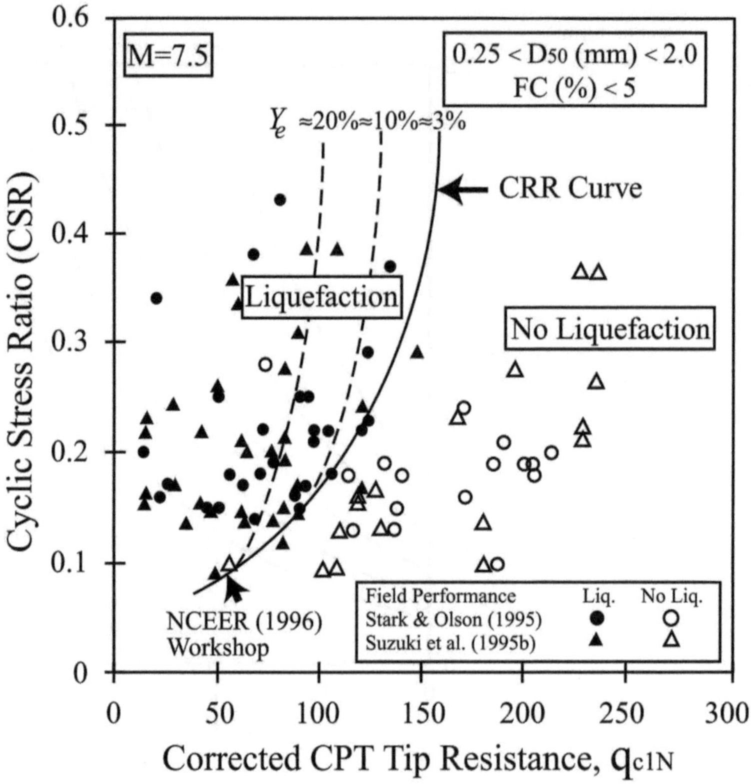

Fig. 6.13 Calculation of CRR from CPT Data along with Empirical Liquefaction Data (after 282)

$\Delta\lambda_{a_i,m_j}$ is the incremental annual frequency of exceedance for a_i and m_j obtained by deaggregating the seismic hazard curve with respect to magnitude and $P[\text{FS} > \text{FS}^*]$ is the conditional probability that obtained factor of safety (FS) will exceed a target value (FS*) as per Eq. (6.11) proposed by Cetin et al. [61].

$$P[\text{FS} > \text{FS}^*|a_i, m_j] = \Phi\left(-\frac{1}{\sigma_\varepsilon}\begin{bmatrix} (N_1)_{60}\,(1+\theta_1 FC) - \theta_2 \ln(\text{CSR}_{eq,i}\text{FS}^*) \\ -\theta_3 \ln(m_j) - \theta_4 \ln(\sigma'_{v0}/P_a) + \theta_5 FC + \theta_6 \end{bmatrix}\right) \tag{6.11}$$

Here FS is the factor of safety and FS* is the target value of factor of safety. $\text{CSR}_{eq,i}$ is the cyclic stress ratio for an acceleration a_i corresponding to site class D obtained using probabilistic hazard analysis and magnitude m_i. $\text{CSR}_{eq,i}$ will be calculated for all the acceleration and magnitude level. θ_1 to θ_6—are the regression coefficient. $(N_1)_{60}$ is the corrected N value (for energy, and overburden pressure). P_a and σ'_{v0} are atmospheric pressure and effective overburden pressure (in the same unit), FC is the fine content in percentage and Φ is the cumulative normal distribution function. From the MARE curve, FS values corresponding to 475 year and 2500 year return period (having 10% and 2% probability of exceedance in 50 years) were evaluated. This FS values were integrated over depth to obtain the LPI for that particular bore hole.

Liquefaction potential can also be characterized in terms of SPT resistance required to prevent liquefaction at a particular depth, N_{req} and $N*_{req}$ are the targeted SPT N values. This method is best suited for assessing liquefaction from meso- to macro-level. It is to be noted that both N_{req} and $N*_{req}$ are the SPT N values corrected for energy, overburden pressure, rod length, sampler, borehole diameter, and percentage of fines. $\lambda_{N*_{req}}$ is the mean annual rate of exceedance with which N_{req} exceeds $N*_{req}$ and is given in Eq. (6.12) [374].

$$\lambda_{N*_{req}} = \sum_{i=1}^{N_a} \sum_{j=1}^{N_m} P[N_{req} > N*_{req} | a_i, m_j] \Delta\lambda_{a_i m_j} \tag{6.12}$$

Here a_i and m_j are the incremental values in earthquake magnitude and acceleration level and the probability of N_{req} exceeding a predefined value of $N*_{req}$ $P\left[N_{req} > N*_{req}\right]$ is given by Eq. (6.13). $\Delta\lambda_{a_i m_j}$ is the incremental annual frequency of exceedance for acceleration a_i and magnitude m_j and $P[N_{req} > N*_{req} | a_i, m_j]$ is given by Eq. (6.14) as per [374]. A mean annual rate of exceedance curve for N_{req} is developed from which SPT N value required to resist liquefaction corresponding to desired return period can be estimated.

$$P[N_{req} > N*_{req} | a_i, m_j] = \Phi\left[-\frac{N*_{req} - \theta_2 \ln(\mathrm{CSR}_{eq,i}) - \theta_3 \ln(m_j) - \theta_4 \ln\left(\sigma'_{vo}/P_a\right) + \theta_6}{\sigma_s} \right] \tag{6.13}$$

Similar to SPT-based liquefaction assessment, CPT-based analysis can also be carried out for any region. The main advantage of using CPT data is that the percentage of errors is considerably less than that of SPT method. The mean annual rate of exceedance $\left(\lambda_{q*_{c1-req}}\right)$ of CPT value (q_c) required to resist liquefaction is given as per Eq. (6.14).

$$\lambda_{q*_{c1-req}} = \sum_{i=1}^{N_a} \sum_{j=1}^{N_m} P\left[q_c > q*_{c1-req} | a_i, m_j\right] \Delta\lambda_{a_i m_j} \tag{6.14}$$

Here $q*_{c1-req}$ is the targeted value; N_m—number of magnitude increments; N_a—number of peak ground acceleration; $\Delta\lambda_{a_i m_j}$ is the incremental annual frequency of exceedance for acceleration a_i and magnitude m_j. Now $P\left[q_{c1} > q*_{c1-req} | a_i, m_j\right]$ is given by Eq. (6.15) as per [225, 375]. From the mean annual exceedance curve, CPT values required to prevent liquefaction for any return period can be estimated.

$$P\left[q_{c1} > q *_{c1-req} \left| a_i, m_j \right.\right]$$
$$= \Phi\left[-\frac{\begin{array}{l}\left(q*_{c1-req}\right)^{1.045} + q *_{c1-req}\left(0.110R_f\right) + 0.001R_f \\ + c\left(1 + 0.850R_f\right) - 7.177\ln(\mathrm{CSR}_{eq,i}) \\ - 0.848\ln m_j - 0.002\ln\left(\sigma'_{v0}\right) - 20.923\end{array}}{1.632}\right] \tag{6.15}$$

6.6 Liquefaction Potential Index

The methods described in the previous sections measure the liquefaction potential at specified depths (the depth selected in the present study was 3 and 6 m). However by using the Liquefaction Potential Index (LPI, proposed by Iwasaki et al. [145]) this shortcoming can be over come. In this method it is assumed that the severity of liquefaction should be proportional to the thickness of the liquefied layer, proximity of the liquefied layer to the surface, and the factor of safety of the liquefied layer. The LPI, which was proposed by Iwasaki et al. [145], is given by:

$$\mathrm{LPI} = \int_0^{20} F_z W(z) dz \tag{6.16}$$

where z is the depth, F_z is a function of factor of safety as per Eq. (6.17) [339], and $W(z)$ depth weightage factor ($W(z) = (10 - 0.5z)$ for $z \leq 20$ and $W(z) = 0$ for $z > 20$).

$$F_z = \begin{cases} 0, & \text{for FS} \geq 1.2 \\ 1 - \text{FS} & \text{for FS} < 0.95 \\ 2 \times 10^6 e^{(-18.427\text{FS})} & \text{for } 1.2 > \text{FS} > 0.95 \end{cases} \tag{6.17}$$

Sonmez [339] proposed a classification of liquefaction potential of a site based on LPI (Table 6.1).

Table 6.1 Liquefaction vulnerability classification based on LPI [339]

LPI range	Liquefaction vulnerability
<2	Low
2 –5	Moderate
5 –15	High
>15	Very high

6.7 Liquefaction Hazard Assessment at Micro-Level: A Case Study of Kalpakkam NPP Site

In engineering design point of view, nuclear power plant structures are meant to survive all possible natural disasters, especially induced effects of earthquakes. Failure of any nuclear power plant structure can result in a cataclysm and the safety of such structures during natural disasters such as earthquakes is of great concern even to the common people. Subsurface investigations conducted in the Kalpakkam NPP site reveal the presence of loose sand layers at shallow depth along with high water table, making the site vulnerable to liquefaction hazard. The liquefaction hazard for the Kalpakkam NPP site (at micro-level) was evaluated in terms of liquefaction potential index (LPI) introduced by Iwasaki et al. [145], which is obtained by integrating factor of safety against liquefaction at all depths. The LPI was estimated for the Kalpakkam NPP site based on field test data, using both deterministic and probabilistic approaches by James et al. [152].

6.7.1 *Deterministic Approach*

In the deterministic methodology, PHA at ground surface level was first evaluated using deterministic seismic hazard analysis (DSHA). Using the surface level PHA value, a simplified methodology proposed by Seed and Idriss [309] was used to estimate cyclic stress ratio (CSR) at various depths of each borehole. Cyclic resistance ratio (CRR) of the soil was computed based on the corrected SPT value $[(N_1)_{60,cs}]$ as per the method suggested by Idriss and Boulanger [131]. For each borehole, the factor of safety at various depths were then evaluated and these values were integrated to obtain the LPI for that particular borehole. Figure 6.14 presents the spatial variation in LPI values from the deterministic methodology for the Kalpakkam NPP site. From Fig. 6.14 it is evident that the northern part of the site is having very high liquefaction hazard with LPI value of 15 and above. Central part of the site has low liquefaction hazard with LPI less than 2.

6.7.2 *Probabilistic Approach*

One of the major shortcomings in the deterministic method is that it considers only one critical magnitude-acceleration scenario for the evaluation of liquefaction potential. In order to overcome this shortcoming, the LPI values were evaluated using a probabilistic performance based methodology proposed by Kramer and Mayfield [182] as well. The probabilistic model captures various uncertainties involved in earthquake process such as the uncertainty in magnitude, location, and the recurrence rate. The mean annual rate of exceedance (MARE) curve, showing

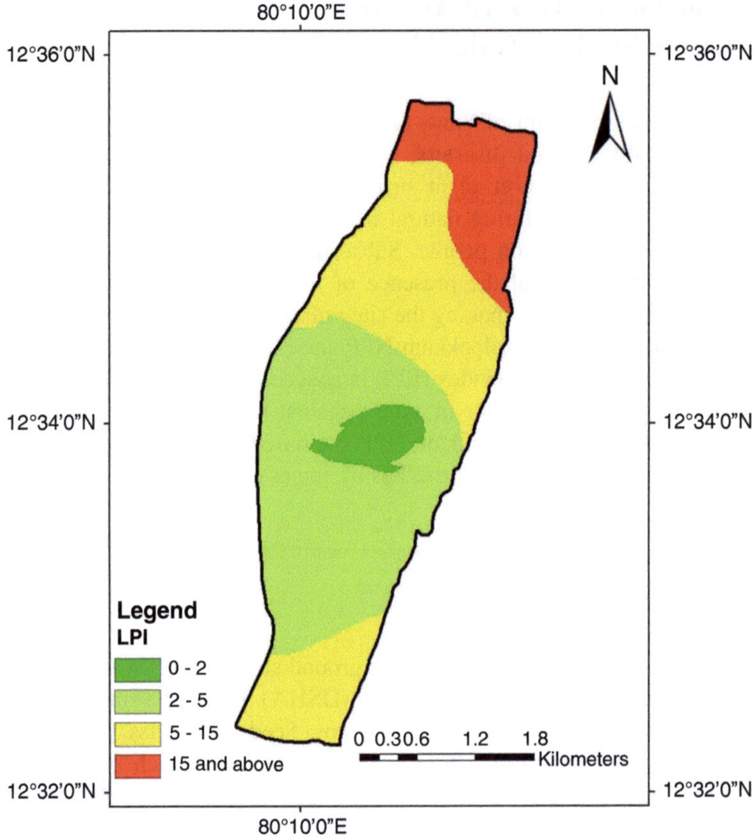

Fig. 6.14 Spatial variation in LPI for the Kalpakkam NPP site obtained from the deterministic method (after [152])

the variation factor of safety (FS) for different return periods was developed [152]. From the MARE curve, FS values corresponding to 475 year and 2500 year return period (having 10% and 2% probability of exceedance in 50 years) were evaluated. A typical MARE curve showing the variation of FS with return period at 3 m depth for a typical borehole is presented in Fig. 6.15. These FS values were integrated over depth to obtain the LPI for that particular bore hole. Figures 6.16 and 6.17 present the spatial variation of LPI value with 475 year and 2500 year return period. Similar to the results obtained from deterministic methodology, probabilistic analysis also shows the presence of highly liquefaction vulnerable region in the northern part of the Kalpakkam NPP site. It is also noted from Figs. 6.16 and 6.17 that the central area of the site has low to moderate hazard value as per the classification provided by Sonmez [339].

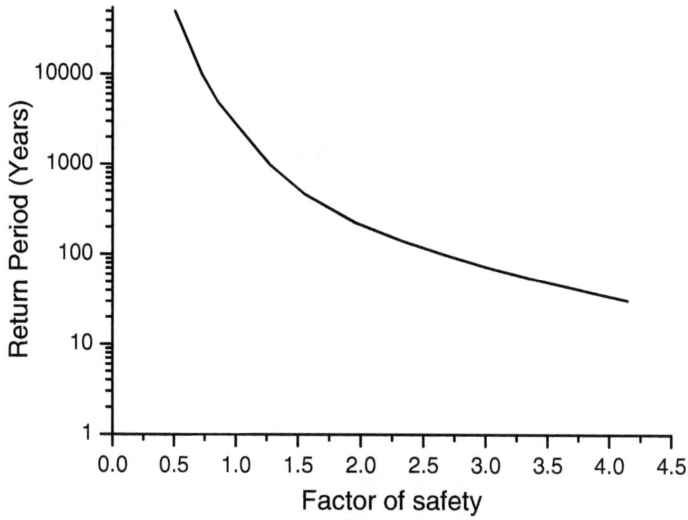

Fig. 6.15 MARE curve showing variation of FS against liquefaction with return period (after [152])

6.8 Liquefaction Hazard Assessment at Macro-Level: A Case Study of Karnataka State

Assessment of liquefaction hazard at macro-level is also essential in order to identify vulnerable regions within the state of Karnataka. However it is very difficult to generate database containing the borelog details for such large region for assessing liquefaction hazard. Hence, the liquefaction hazard must be estimated in terms of parameter which can be easily obtained for any given location using most common geotechnical tests. The liquefaction hazard for the state of Karnataka is estimated in terms of SPT and CPT values required to resist liquefaction by [150]. The standard penetration test (SPT) is the most widely used geotechnical field tests, especially in India for sub-surface exploration. In this study, a probabilistic methodology presented by Kramer and Mayfield [182] is adopted in order to predict the liquefaction hazard for the state of Karnataka in terms of SPT and CPT values at 3 m depth (considering to be the probable depth of a shallow foundation). The state of Karnataka was divided into small grids of 0.05° × 0.05° (approx. 5 km × 5 km) and seismic site characterization of each grid area was carried out using the topographic slope map as mentioned in this chapter. The liquefaction hazard was estimated for the grid points belonging to site class D. Spatial variations in the SPT and CPT values required to resist liquefaction for a definite return period (475 year and 2500 year), throughout the Karnataka state, were evaluated and are presented.

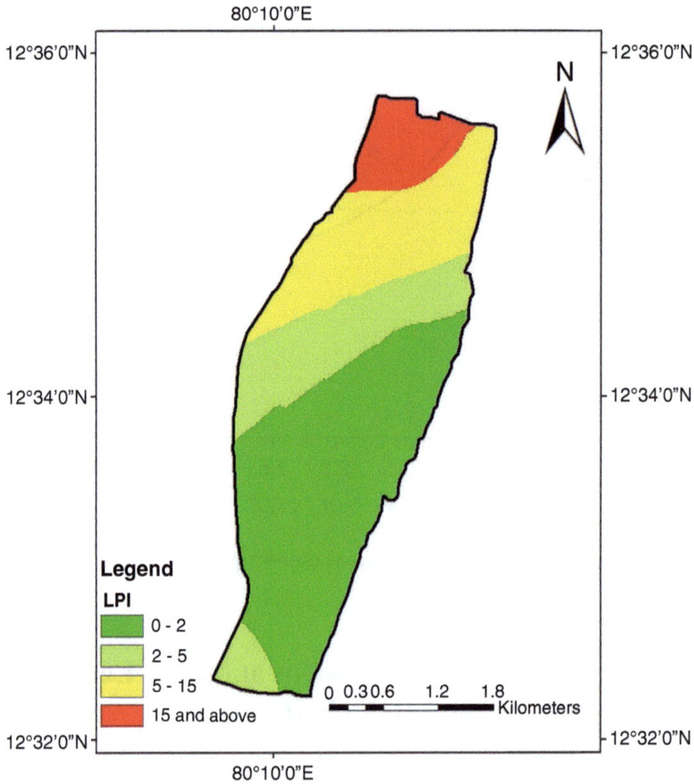

Fig. 6.16 Spatial variation in LPI values for Kalpakkam NPP site corresponding to return period of 475 years (after [152])

6.8.1 Prediction of SPT-N value Required to Resist Liquefaction

In India, engineers still depend on SPT for the site investigation, prior to any construction project and hence the estimation of liquefaction hazard in terms of SPT-N value will be most advantageous.

Figures 6.18 and 6.19 present the spatial variations in the SPT values required to resist liquefaction at 3 m depth, for 475 year and 2500 year return period, throughout the Karnataka state. From Figs. 6.18 and 6.19, it is evident that the southern (regions close to Bangalore–Mysore) and the northern parts (region close to Bidar district) are expected to have high liquefaction hazard. The SPT value required to prevent liquefaction ranges from 10 to 20, for a return period of 475 years and 20 to 30, considering a return period of 2500 year. That is, any given site in this region has less chance of liquefy, if the actual SPT value at 3 m depth is more than the predicted SPT value corresponding to the respective return period.

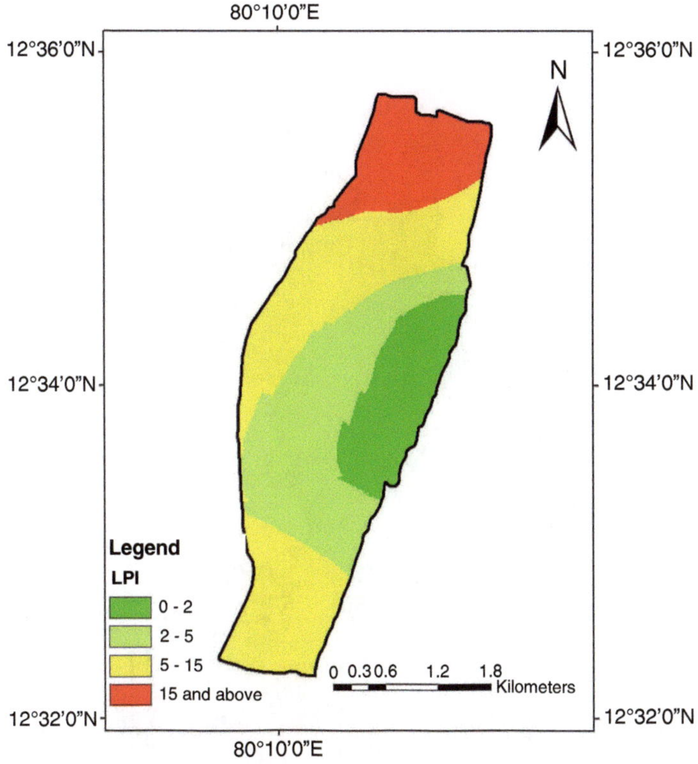

Fig. 6.17 Spatial variation in LPI values for Kalpakkam NPP site corresponding to return period of 2500 years(after [152])

Considering the return period of 475 years, the other parts of Karnataka state have low to moderate liquefaction hazard where SPT values required is less than 10. Similarly considering a return period of 2500 year, the SPT values required to prevent liquefaction is less than 20 for the interior regions of the Karnataka state. Kaiga, where a nuclear power plant site is located is found to have a very low liquefaction hazard. The SPT value required to resist liquefaction for this region is less than 5 considering both 475 year and 2500 year return period.

6.8.2 Prediction of CPT Value Required to Resist Liquefaction

In spite of being the most popular geotechnical field test for the site investigation, the SPT contains a lot of errors and thus require various corrections. Hence the liquefaction hazard for the region was also expressed in terms of CPT value as the percentage of errors is considerably less in the CPT compared to SPT

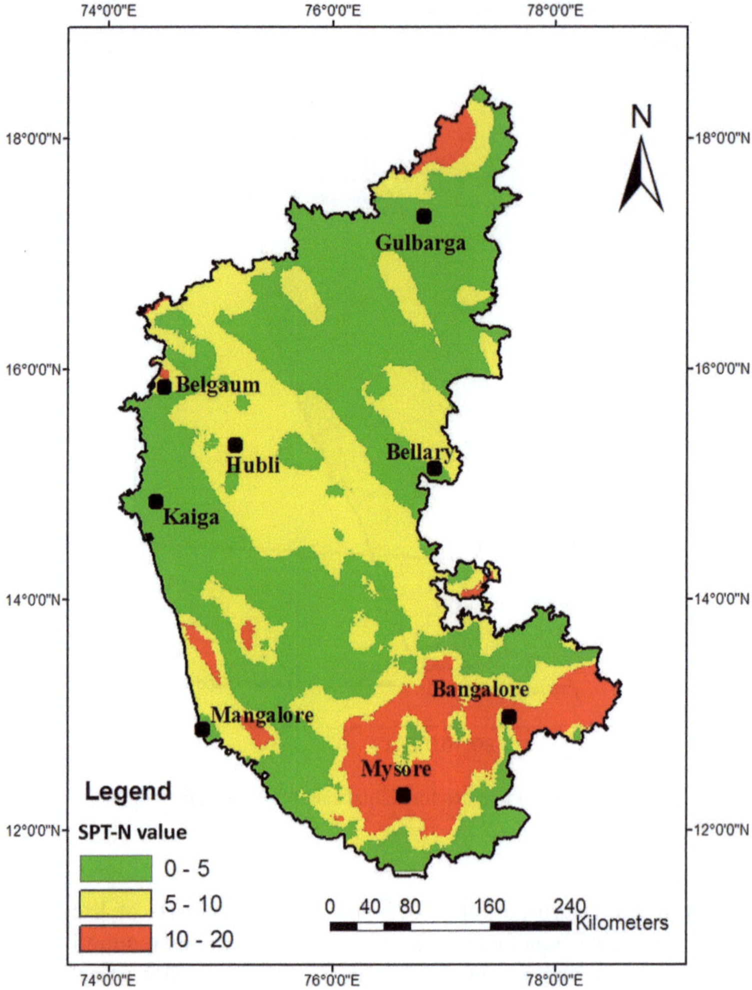

Fig. 6.18 Spatial variation in SPT values required to resist liquefaction at 3 m depth throughout the Karnataka state for 475 year return period (after [150])

method. The probabilistic methodology for the estimation of liquefaction hazard in terms of CPT values required to resist liquefaction is described in Sect. 6.5.3. Figures 6.20 and 6.21 present the spatial variations in the CPT values required to resist liquefaction at 3 m depth, for 475 year and 2500 year return period, throughout the Karnataka state. As evident from the above study (Figs. 6.20 and 6.21) that the southern (regions close to Bangalore–Mysore) and the northern parts (region close to Bidar district) of the state are expected to have a high liquefaction hazard. For both the regions, the CPT value required to resist liquefaction ranges from

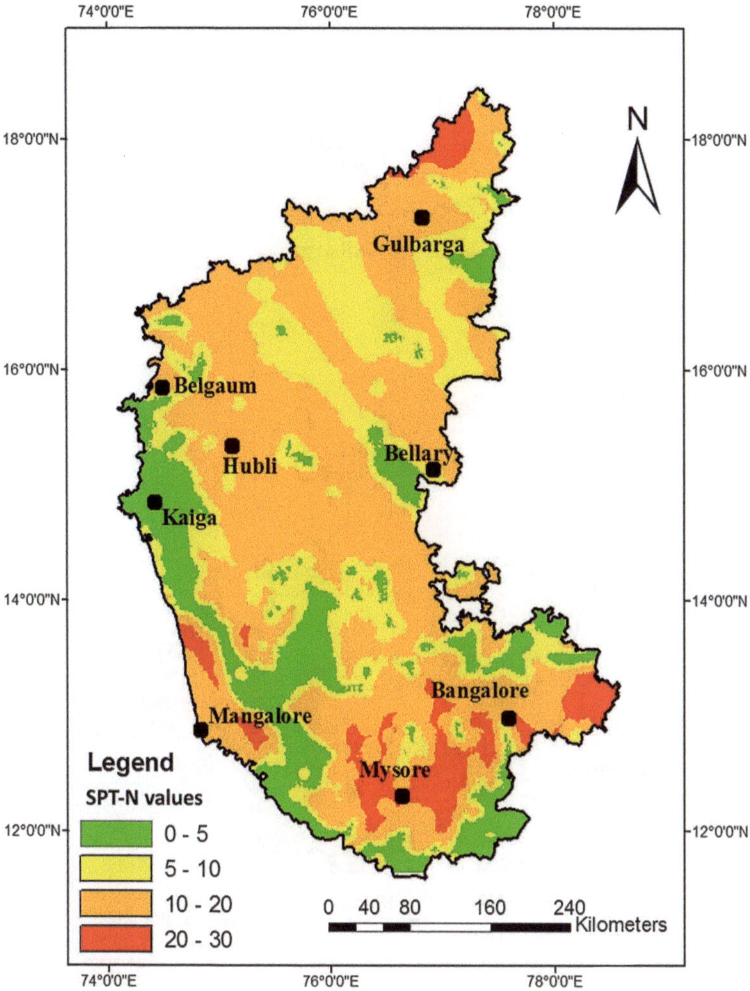

Fig. 6.19 Spatial variation in SPT values required to resist liquefaction at 3 m depth throughout the Karnataka state for 2500 year return period (after [150, 151])

7 to 13 MPa, considering both 475 year and 2500 year return period. The CPT values required to resist liquefaction throughout the interior parts of Karnataka, for a return period of 475 year is found to be in the range of 2–7 MPa. However, for a return period of 2500 year the CPT value required to resist liquefaction ranges from 5 to 13 MPa. Hence the interior region of the Karnataka state is found to have moderate to high liquefaction hazard. Such a high value of liquefaction hazard is due to high surface level PHA, as only linear source is considered.

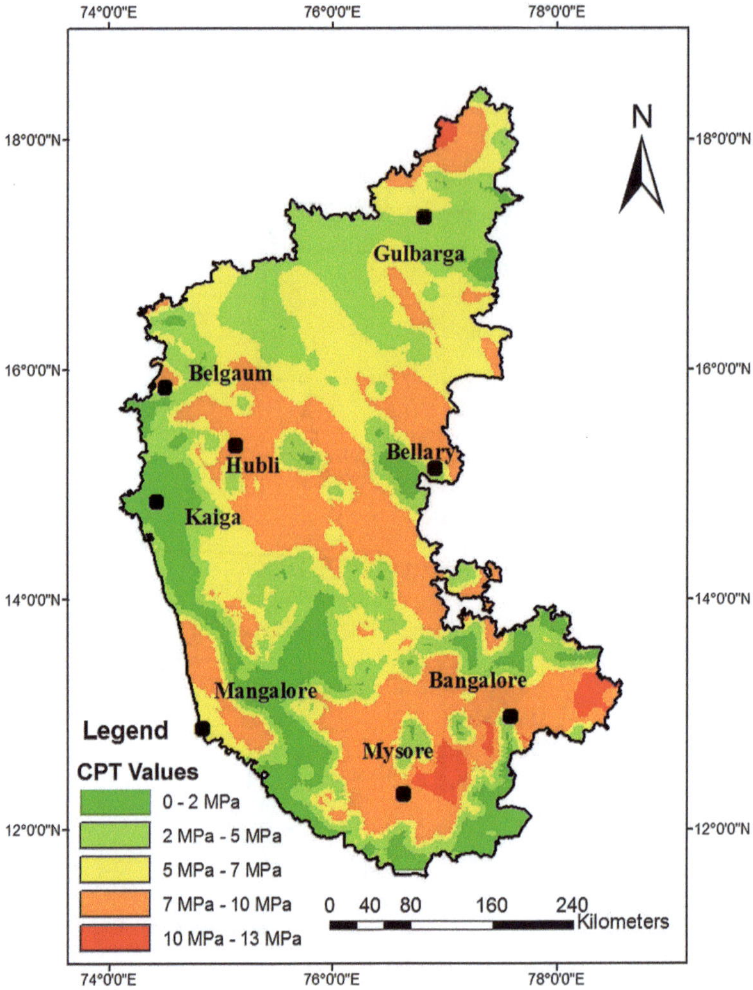

Fig. 6.20 Spatial variation in CPT values required to resist liquefaction at 3 m depth for 475 year return period (after [150])

6.9 Liquefaction Mitigation

Soil liquefaction is one of the worst seismic hazards that can occur during an earthquake. Several studies in the past have shown that the earthquake induced liquefactions of soils can cause significant damages to the man-made infrastructures. Possible causes, factors affecting the soil liquefaction along with its mechanism and evaluation procedure have been discussed in the above sections of this chapter. This section will discuss about the different methodologies developed to reduce or eliminate damage due to liquefaction. In recent years, several techniques have been

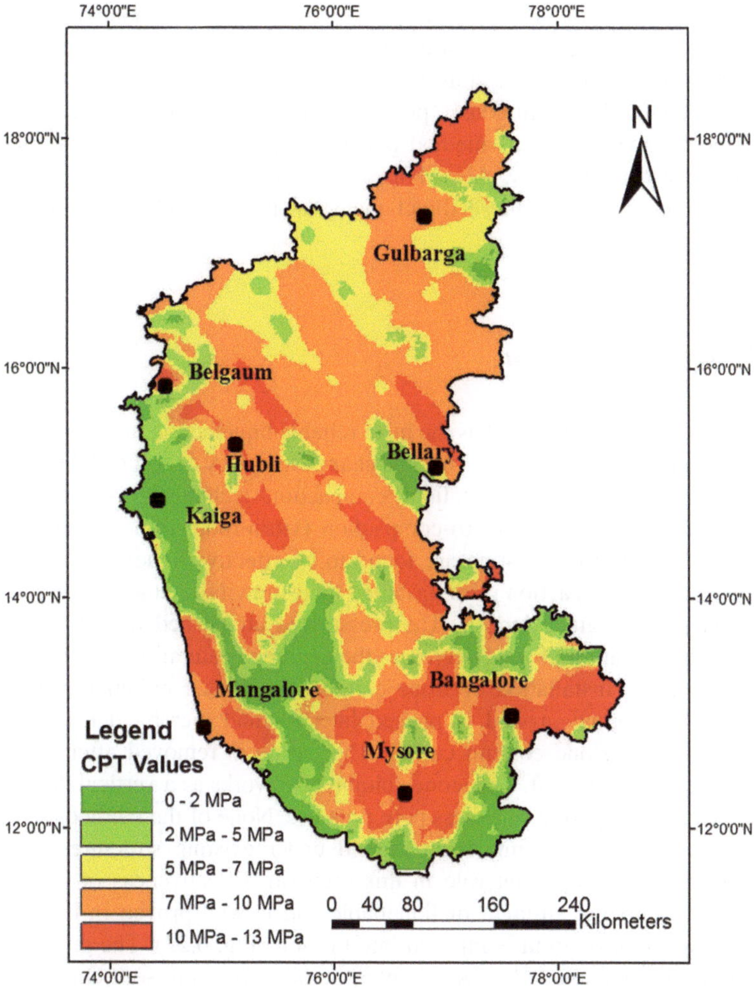

Fig. 6.21 Spatial variation in CPT values required to resist liquefaction at 3 m depth for 2500 year return period (after [150])

developed by many researchers toward liquefaction mitigation techniques, for the new as well as the existing structures. These techniques can be broadly classified into following.

6.9.1 Direct Methods: Soil Replacement and Dewatering

Soil replacement method involves replacement of liquefiable soil with a well-graded fill which has high resistance to liquefaction. Even though the method is very reliable, it becomes difficult and expensive if liquefiable soil is found for significant depth. Also, it is inapplicable if water table lies at shallow depth. Generation of excess pore water pressure leading to liquefaction can be prevented by lowering groundwater table by continuously pumping out water from the zone of interest. The biggest drawback of this method is that pumping is to be done permanently to keep the water table low. Also, both upstream and downstream seepage cut-offs are required. Hence, these direct methods are not preferable for practical applications.

6.9.2 By Densification

Soil densification is considered as the most reliable liquefaction mitigation method. Densification methods reduce the volume change tendency of soil and thus decrease the liquefaction potential. The field compaction techniques like compaction piles, dynamic compaction, vibro-compaction (vibro-rod, vibro-floatation, vibro-replacement, etc.), deep blasting are found to be effective liquefaction mitigation techniques. Sand compaction piles, installed by using vibro-hammers, displace and thus densify the original deposit. The pressurized air pushed into the ground in order to help migration of sand decreases the degree of saturation and thus provide an added benefit. Installation of stone columns is another effective remediation measure. The original ground is pushed laterally and thereby densified during the installation of stone column casing. The casing is removed after filling and compacting the stones in it. The stone column also works as a vertical drain which accelerates the dissipation of excess pore pressure. None of the methods discussed above are applicable for improving ground under existing structures. Grouting technique plays an important role in this scenario. Effectiveness of compaction grouting in improving strength of liquefiable soil is well proven in history [47]. A thick, low mobility grout is injected into the soil (Fig. 6.22). The grout remains as a homogeneous mass and does not enter the pore spaces in soil. Once the grout expands, the soil particles get displaced and thus the soil is densified. The grout mix used for densification purpose generally consists of silty sand, fly ash, cement, and water [13]. Clay content is restricted in grout mix to attain proper compaction and to avoid hydraulic fracturing. The pressure and rate of injection should be maintained such that heaving of overlying ground does not occur (Fig. 6.22).

Mitsunan et al. [219] proposed the concept of superlime piles which use burnt lime for densification of loose sand. Superlime piles are bored piles filled with a mixture of burnt lime, gypsum, blast furnace slag, and sand. The mixture absorbs groundwater and swells considerably such that the surrounding soil gets densified.

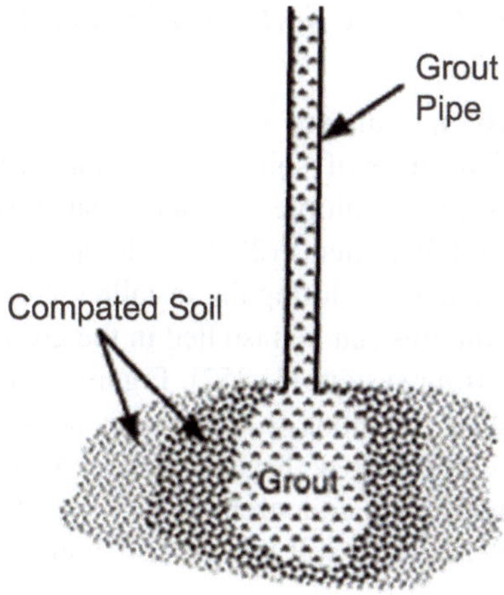

Fig. 6.22 Soil Densification by compaction grouting (after [13])

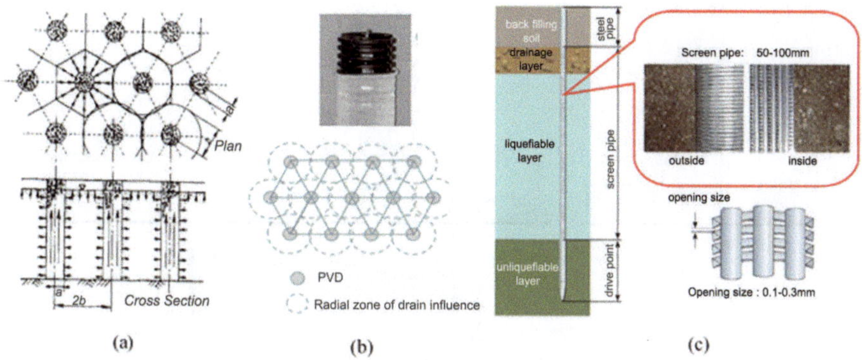

Fig. 6.23 (**a**) Arrangement of gravel drain system (after [390]) (**b**) PVD and an example of a triangular installation grid (after [127]) (**c**) Concept of screen pipe (after [119])

6.9.3 By Mechanical Intrusions for Drainage

Loss of shear strength can be reduced by accelerating dissipation of excess pore water pressure with the help of gravel drains [129, 300, 305, 390]. The vibrations produced during installation of gravel drains also help in densification of surrounding soil [1]. Prefabricated vertical drains (PVD) are also being used for providing drainage as an alternative to conventional gravel drains [127]. The shortening of drainage path inhibits rise in excess pore pressure and thus prevents liquefaction. Harada et al. [119] used screen pipes as vertical drains which can be easily installed in places with limited space. Figure shows a typical screen pipe. Different drainage schemes are shown in Fig. 6.23.

6.9.4 By Soil Reinforcements

By introducing elements which can carry more stress than soil, deformation resistance of soil can be enhanced. Installation of stone columns in the loose soil has multiple advantages such as: (1) it provides deformation restricting effect (reinforcement) (2) it accelerates pore pressure dissipation (drainage) (3) soil gets compacted during the installation of stone column casing (densification) [1]. Stone columns can be installed in the ground by vibro-replacement [1, 49] or using auger casing system [1, 252]. Figure 6.24 shows the installation of stone columns using these methods. Liquefaction resistance of natural deposits can be enhanced by reinforcing with geosynthetic or natural fibers [183, 203, 221]. Another way of improving soil strength is by incorporating structural elements such as micropiles [218, 228], soil nailing [217], etc. In the recent past, several studies were carried

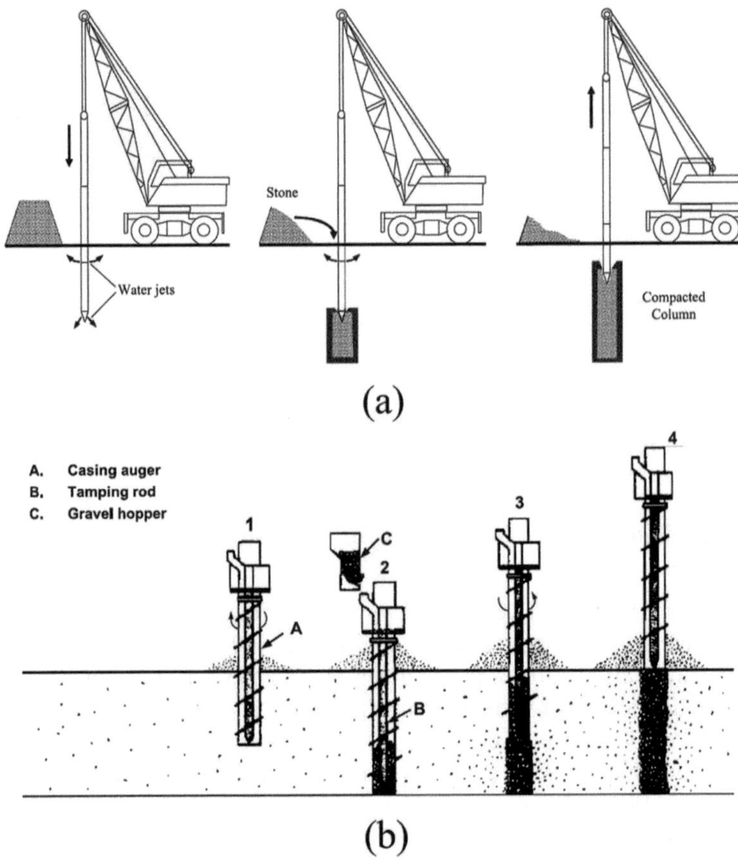

Fig. 6.24 Gravel drain installation (**a**) by vibro-replacement (after [1]) (**b**) by auger-casing system with internal gravel feeding compaction-rod system (after [340])

out toward the use of scrap tyre chips for the liquefaction mitigation in sands [210, 271, 365]. Adding tire chips to sand reduces its modulus of rigidity. Under the seismic shear force, the mixture can undergo volume contraction which tends to decrease the excess pore pressure [360]. Tire chips can also be used as a drain to dissipate the excess pore pressure rapidly [394]. Moreover, the geotechnical application of these tire chips provides a sustainable and eco-friendly solution to the problems related to the recycling of used tires.

6.9.5 By Chemical Modifications: Solidification

In this method, soils cohesive strength is improved by mixing cementitious material with it and thus prevents the soil particle movement. When injected, these grouting fluids permeates into the soil voids due its low viscosity without disturbing the soil structure. This technique is termed as permeation grouting. It strengthens the ground by cementing soil particles together. Also, it reduces the water flow by plugging soil pores. Sodium silicate grouts consisting of sodium silicates, acrylamides, lignosulfonates, and resins [168] are the most popular ones. In jet grouting, the mixing of stabilizer with the soil is accomplished by high pressure fluid jets. The grout is generally a water-cement mixture (Fig. 6.25). A small hole is drilled to the target depth and grout is jetted into the soil through the nozzles near the tip of drill rod. The rod is kept rotating during the withdrawal and the jetting is continuously done from bottom to top of the drill hole. The mechanical mixing and sometimes partial replacement of soil with cementitious materials using mixing augers is termed as in situ soil mixing. The procedure includes penetration of augers into the target depth, pumping of stabilizer through shaft and mixing of stabilizer with the soil by the rotation of

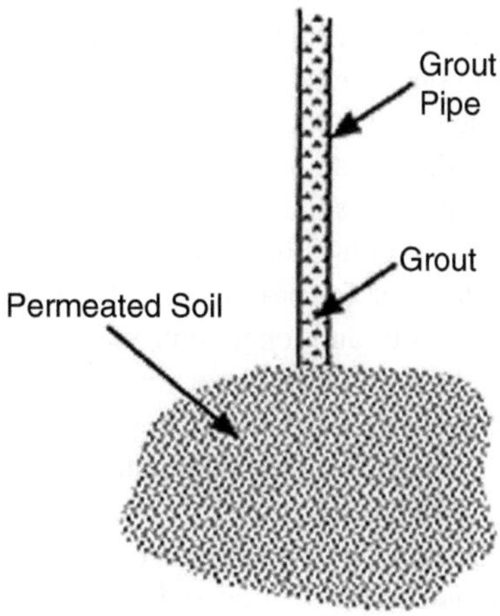

Fig. 6.25 Soil solidification by permeation grouting (after [13])

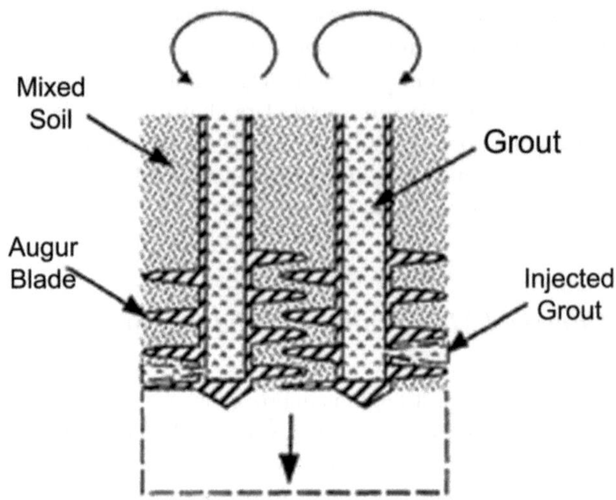

Fig. 6.26 In situ soil mixing technique (after [13])

auger (Fig. 6.26). During the withdrawal of augers a second stage of mixing is performed. As a result, strength of the mixed zone is improved and its permeability is reduced.

6.9.6 Passive Site Stabilization Techniques

In traditional methods of liquefaction mitigation, the size of treated area is limited. Vibration produced during construction creates disturbance to existing structures [128]. To overcome these limitations, Gallagher [98] proposed the concept of passive site stabilization method. In this method, the stabilizers of low viscosity are supplied to the ground through trenches and carried by natural groundwater flow to the target area. The gel time of stabilizing material should be long and controllable so that it can have enough time to flow slowly to the zone of interest. The concept of Passive site remediation is illustrated in Fig. 6.27.

Colloidal silica, which is an aqueous suspension of silica particles, is an effective stabilizer for passive site stabilization. Low concentration colloidal silica has low viscosity and long, controllable gel time [99]. On gelation it transforms into silica gel, thereby creating a very dense matrix with loose sand particles which possess an improved resistance to liquefaction [100]. Therefore this method can be categorized as a solidification technique.

Haldavnekar et al. [116] experimentally proved that addition of bentonite boosts the liquefaction resistance of sand. Addition of bentonite converts the pore fluid to concentrated bentonite gel which exhibits elastic behavior for a wide range of strain.

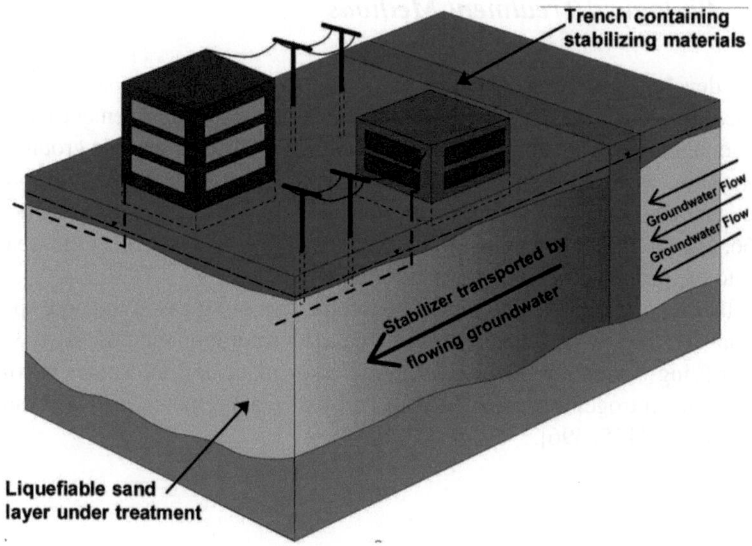

Fig. 6.27 Passive site remediation for mitigation of liquefaction risk (after [99])

The enhanced cyclic resistance is attributed to this extended elastic response of pore fluid [81, 82]. The rheological properties of bentonite suspension can be controlled by using sodium pyrophosphate (SPP). Rugg et al. [293], by examining undrained shear response of sand-bentonite mixture, observed that the mixture possesses no cohesion and its friction angle is same as that of untreated sand. It shows that the soil structure is preserved even after the permeation of bentonite.

6.9.7 By Inducing Desaturation

In recent past, desaturation has emerged as an effective technique for the liquefaction mitigation. In this method, the liquefaction resistance of the soil increased by reducing its degree of saturation by injecting air or biogas [128]. Several studies were carried out toward liquefaction mitigation by desaturation [141, 230, 249, 364, 393, 396]. The safety against liquefaction of soil can be assured by decreasing its degree of saturation [26]. A small reduction in the degree of water saturation of sand soil can result in a significant increase in shear strength against liquefaction [85]. Nakazawa et al. [230] have shown that cyclic strength in laboratory test specimens can be more than twice as high in partially saturated soil than fully saturated soil. By injecting air into the site with the help of special equipments, degree of saturation can be reduced to desired level [53, 85, 249]. Once the air injection is done for a short time period, the resulting desaturation can last for years [249].

6.9.8 Biological Treatment Methods

By the development of a new discipline called Microbial Geotechniques, the biological treatment methods evolved as effective ground improvement technique. When some microorganisms are injected along with nutrients into the ground under low pressure, as a result of some chemical reactions, products having ability to cement the sand particles are generated. This increases thc shear strength of the soil [66, 76, 372]. The slow rate of chemical reactions helps to inject the treatment fluid into the ground efficiently before gelling.

Another type of biological treatment is production of biogas in the pore space by microbial activities which can be considered as a desaturation method. By the action of denitrifying bacteria, reduction of nitrate takes place and molecular nitrogen is released. This nitrogen occupies pore space by displacing pore water and thus soil get desaturated [125, 396].

Chapter 7
Principles and Practices of Seismic Zonation

7.1 Introduction

Assessment of various earthquake hazards such as the ground shaking, site amplification and liquefaction, both at micro and macro-levels, has been discussed in previous chapters. The main objective of this chapter is to describe the seismic zonation at the micro as well as macro-level, based on the spatial variation of these hazards. A detailed analysis of the spatial distribution of these hazards reveals that the distribution pattern of each hazard in a given study area differs from the other. It is worth noting that a region with a high level of shaking at the bedrock level need not exhibit high site amplification or liquefaction hazard. So it is difficult to carry out the seismic zonation if each of the earthquake hazards is treated separately. Hence there is a need to develop an index number which can effectively represent the combined effect of all these hazards at a particular location.

7.2 Integration of Various Hazards

The estimation procedures of major hazards associated with earthquakes such as ground shaking, liquefaction, landslide, etc. were discussed in the previous chapters of this book. As each of these hazards is predominantly influenced by local geological setting rather than source characteristics of the earthquake, the spatial distribution of intensity of each of these in a given region is distinct from the other. Thus it is difficult to assess the risk and the vulnerability of a region when these hazards are treated separately. Hence all major earthquake hazards are to be integrated so that their combined effect can be understood in a better way. This chapter describes the methodology for integrating various hazards associated with earthquakes. All the major seismic hazards for the two study areas are integrated to a hazard index value using the Analytical Hierarchy Process (AHP) proposed

T. G. Sitharam et al., *Comprehensive Seismic Zonation Schemes for Regions at Different Scales*, https://doi.org/10.1007/978-3-319-89659-5_7

by Saaty [295]. The hazard index effectively represents the consolidated effect of all major seismic hazards and hazard influencing parameters. The region was then demarcated based on the spatial variation of hazard index value.

The first step of hazard integration is to prioritize various themes (here seismic hazards). Hierarchical arrangement of various themes is subjected to user judgement. After prioritizing, the weightage is assigned each theme. The theme having high priority is assigned with higher weightage. After establishing hierarchy among various themes, normalized weights for each theme is then calculated using a pair-wise comparison matrix. In a pair-wise comparison matrix each theme was compared with the other as the ratio between their respective weights. The normalized weight for each theme is then calculated by averaging the values in each row of the matrix with the total weight. A second level of categorization was made within each of the themes and a rank was assigned to each category. Range of each category and number of categories within a particular theme are also based on user judgement. These ranks (for each theme category) are then normalized using Eq. (7.1).

$$X_i = \frac{R_i - R_{\min}}{R_{\max} - R_{\min}} \tag{7.1}$$

Here, R_i is the rank assigned to each category of a single theme, $R_{\min}$ and $R_{\max}$ is the minimum and the maximum rating value of that theme. Hazard index values are then estimated based on normalized weights and ranks by integrating of all themes using the following Eq. (7.2).

$$\mathrm{HI} = \frac{\sum[(\mathrm{EHP}_w)_i(\mathrm{EHP}_r)_i]}{\sum w_i} \tag{7.2}$$

where HI is the hazard index, EHP_w and EHP_r are the normalized weight and rank of ith earthquake hazard parameter in a pair-wise comparison matrix, 'w' is the normalized weight of the theme. Procedures for estimating each of these earthquake hazard parameters are mentioned in the above sections of this chapter.

7.3 Seismic Zonation Studies at Micro-Level

7.3.1 Seismic Zonation of Kalpakkam NPP Site

The hazard integration at the micro-level was carried out for the Kalpakkam nuclear power plant site by James et al. [153]. Since nuclear power plants are considered as sensitive and critical installations with different components founded at different depth, seismic microzonation for these sites is of great importance. Seismic microzonation maps of the Kalpakkam NPP site were prepared based on the

Table 7.1 Parameter and weights used for generating hazard index map at the micro-level (after [153])

Earthquake hazard parameter (EHP)	Weights
Peak horizontal acceleration (PHA)	6
Amplification factor (Amp.factor)	5
Factor of safety against liquefaction (Lq.Fs)	4
Predominant frequency (PF)	3
Average shear wave velocity for top 30 m (V_{s30})	2
Overburden soil thickness/rock depth (Obr.thick)	1

spatial variation of hazard index value. These maps can be selected as a criterion for site selection for various nuclear infrastructures and also prove helpful in assessing the vulnerability for the existing facilities. Major seismic hazards considered in the evaluation of hazard index at the micro-level are (1) intensity of ground shaking at bedrock (in terms of PHA), (2) site amplification, (3) liquefaction potential, and (4) the predominant frequency of the earthquake motion at the surface. Apart from these earthquake hazards, other significant parameters representing local geology such as the average shear wave velocity for top 30 m (V_{s30}) and the thickness of overburden soil (or rock depth) were also considered in the evaluation of hazard index. A list of earthquake hazard parameters (EHP) and corresponding weights used in this study is presented in Table 7.1.

The peak horizontal acceleration (PHA) at bedrock was given highest weightage as it is the most fundamental. Subsequent priorities were allotted to hazards such as amplification factor, factor of safety against liquefaction, and predominant frequency. Other parameters such as average shear wave velocity of the top 30 m overburden (V_{s30}) and thickness of overburden soil (or rock depth) were given least weightages of all. In AHP, a matrix was constructed (Table 7.2) showing the pairwise comparisons (ratios) between the factors of earthquake hazard parameters (EHP). The pairwise comparisons will result in a normalized weight for each parameter and the sum of the normalized weights is unity. Each EHP was categorized into various ranges and for each category a rating or ranking was assigned. Table 7.3 presents the details of each EHP category and the ranks associated with each of them. These ranks (for each EHP category) were then normalized using Eq. (7.1). Once EHPs were identified and prioritized, they were integrated to a hazard index value that ranges from 0 to 1. After the estimation of normalized weight and rating for each category of earthquake hazard parameter, the hazard index for the study area was estimated using Eq. (7.3).

$$\mathrm{HI} = \left(\frac{1}{\sum w}\right)\left[\begin{array}{c}\mathrm{PHA}_w \times \mathrm{PHA}_r + \mathrm{Amp.factor}_w \times \mathrm{Amp.factor}_r \\ +\mathrm{Lq.Fs}_w \times \mathrm{Lq.Fs}_w + \mathrm{PF}_w \times \mathrm{PF}_r + V_{s30w} \times V_{s30r}\end{array}\right] \tag{7.3}$$

where HI is the hazard index, 'w' is the normalized weight of the earthquake hazard parameter (EHP), and 'r' is the normalized rank of a category in the

Table 7.2 Normalized weights for each earthquake hazard parameter (EHP) (after [153])

EHP	PHA	Amp. factor	Obr.thick	V_{s30}	Lq.Fs	PF	Weights
PHA	6/6	6/5	6/4	6/3	6/2	6/1	0.285714
Amp.factor	5/6	5/5	5/4	5/3	5/2	5/1	0.238095
Lq.Fs	4/6	4/5	4/4	4/3	4/2	4/1	0.190476
PF	3/6	3/5	3/4	3/3	3/2	3/1	0.142857
V_{s30}	2/6	2/5	2/4	2/3	2/2	2/1	0.095238
Obr.thick	1/6	1/5	1/4	1/3	1/2	1/1	0.047619

Table 7.3 Normalized rating for each EHP category for micro-level hazard integration (after [153])

EHP	Value range	Weight	Rank	Normalized rank
PHA(g)	≤0.05	0.2875	1	0
	0.05–0.09		2	0.33
	0.09–0.13		2	0.67
	≥0.13		3	1
Amp. factor	1.0–2.0	0.2381	1	0
	2.0–3.0		2	0.5
	≥3.0		3	1
Lq.Fs	≤1	0.1905	3	1
	1.0–2.0		2	0.5
	≥2.0		3	0
V_{s30} (m/s)	≤100	0.0952	6	1
	100–200		5	0.80
	200–300		4	0.60
	300–400		3	0.40
	400–500		2	0.20
	≥500		1	0
Obr.thick (m)	≤5	0.0476	1	0
	5–10		2	0.25
	10–15		3	0.50
	15–20		4	0.75
	>20		5	1

EHP. Deterministic and probabilistic seismic microzonation maps of the study area, showing the spatial variation in the hazard index values were prepared and presented in this chapter.

Figure 7.1 presents the deterministic seismic microzonation map and Figs. 7.2 and 7.3 present the probabilistic seismic microzonation map corresponding to 475 year and 2500 year return period, respectively. The hazard index for preparing the deterministic microzonation map was evaluated using EHP obtained from deterministic methodology such as PHA and factor of safety against liquefaction.

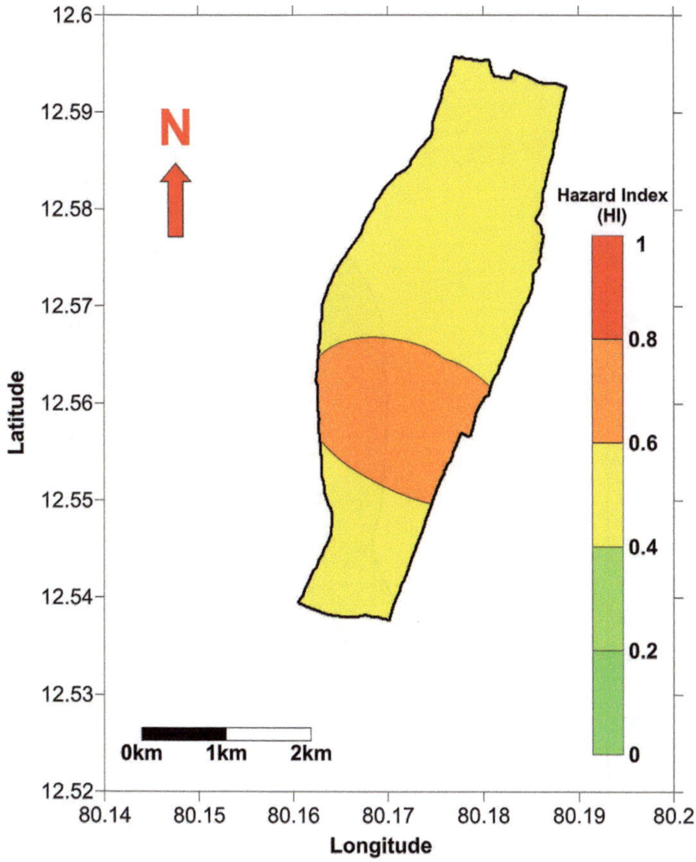

Fig. 7.1 Deterministic seismic microzonation map for the Kalpakkam NPP site (after [153])

Similarly for preparing the probabilistic hazard index map, EHPs such as PHA and factor of safety against liquefaction corresponding to a return period of 475 years and 2500 years were selected.

The deterministic microzonation map (Fig. 7.1) shows that the central sector of the Kalpakkam NPP site close to S-3 and S-4 locations has the higher hazard index (HI) value ranging from 0.6 to 0.8. Such a high value of hazard index is mainly contributed by EHPs such as PHA, amplification factor, and predominant frequency. The deterministic hazard index for the rest of the area in the Kalpakkam NPP site varies from 0.4 to 0.6. Probabilistic microzonation map for 2500 year return period also presents a similar trend in the spatial distribution of hazard index except in the northern sector of the study area, where hazard index value ranges from 0.6 to 0.8. This disparity can be attributed to the fact that the PHA for the northern sector corresponding to 2500 year return period is marginally higher than the PHA obtained from the deterministic methodology. Probabilistic microzonation map for

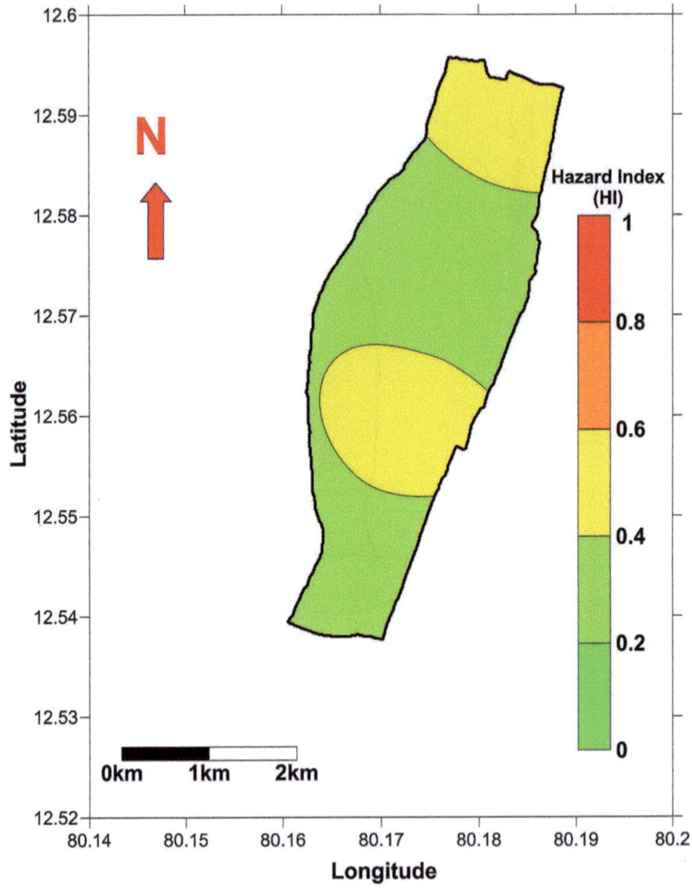

Fig. 7.2 Probabilistic seismic microzonation map of the NPP site (475 year return period after [153])

475 year return period shows that the hazard index value for a major portion of the Kalpakkam NPP site varies from 0.2 to 0.4, except for the central and northern sectors where HI varies between 0.4 and 0.6.

Hence considering the worst scenario from above three maps, the central and northern sectors can be categorized to region having high hazard as per [235], where hazard index varies from 0.6 to 0.8. The rest of the area of the Kalpakkam NPP site can be categorized as a region having moderately high hazard, where hazard index value varies from 0.4 to 0.6.

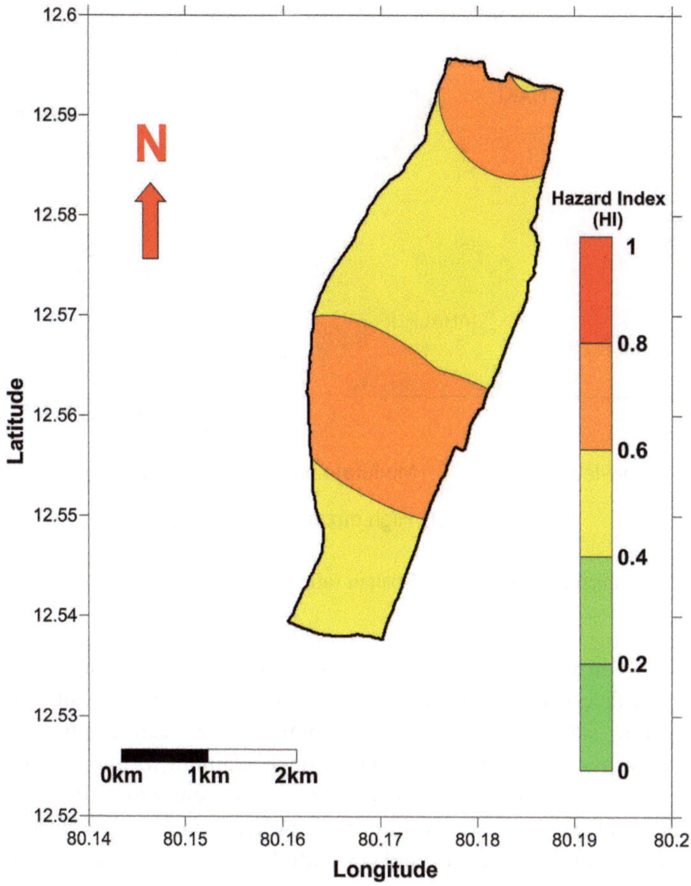

Fig. 7.3 Probabilistic seismic microzonation map of the NPP site (2500 year return period after [153])

7.4 Seismic Zonation Studies at Meso-Level

7.4.1 Seismic Zonation of Jabalpur

Jabalpur microzonation [263] was one of the earliest works in India, towards seismic microzonation with the support from the Department of Science and Technology (DST) New Delhi, Geological survey of India (GSI), Central Region Nagpur, Indian Meteorology Department (IMD) New Delhi, National Geophysical Research Institute (NGRI) Hyderabad, Central Building Research Institute (CBRI) Roorkee, and Government Engineering College, Jabalpur. Seismic hazard in terms of peak ground acceleration (PGA) was estimated at bedrock level using deterministic approach. Site characterization for the entire city was carried out based on geology, geotechnical, and geophysical investigations.

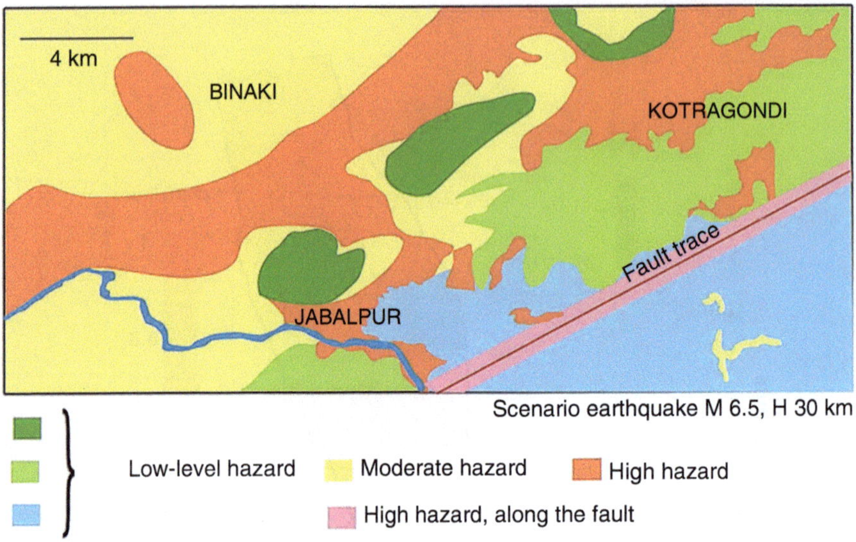

Fig. 7.4 Seismic microzonation map of Jabalpur (after [263])

Based on 30 m equivalent shear wave velocity, measured using multichannel analysis of surface wave (MASW) tests, the entire Jabalpur was categorized into various seismic site class. Microtremor studies have been conducted and the seismic site response was evaluated based on Nakamura technique and receiver function type studies. The liquefaction hazard assessment was estimated based on geotechnical field test data using a simplified approach proposed by Seed and Idriss [309]. The predominant frequency and peak amplification maps were developed and presented. Figure 7.4 presents the seismic microzonation map prepared for Jabalpur.

7.4.2 Seismic Zonation of Delhi

Many researchers [18, 146, 223, 227, 261, 277] have carried out microzonation work for Delhi region. Iyengar and Ghosh [146] evaluated PGA at bedrock level for Delhi using both deterministic and probabilistic approaches. All seismic sources and earthquake events within 300 km from the boundary of Delhi region were compiled for seismic hazard assessment. A quantitative seismic hazard map for the entire Delhi was prepared by considering a grid size of 1 km × 1 km. They have also carried out detailed 1D site response analysis based on the borelog data using SHAKE-91. Parvez et al. [261] has computed the site specific seismic ground motion in a part of Delhi City using a hybrid technique. The hybrid technique was based on the model summation and the finite-difference scheme.

Rao and Neelima Satyam [277] have generated a seismic hazard map at bedrock level for entire Delhi based on finite fault simulation technique using FINSIM computer code. Extensive geotechnical and geophysical tests were conducted and seismic site characterization was also carried out based on V_{s30} value. Microtremor studies have been carried out to compute the seismic site response. They have also carried out liquefaction assessment for Delhi based on SPT data. A first order seismic microzonation map of Delhi was prepared [304]. First they generated five thematic layers viz., Peak Ground Acceleration (PGA) contour, different soil types at 6 m depth, geology, groundwater fluctuation, and bedrock depth. The highest priority of 0.333 was given to PGA and subsequent priorities of 0.266, 0.2, 0.133, and 0.066 were given to soil type, geology, groundwater, and bedrock depth, respectively. These maps were then integrated using Analytical Hierarchy Process (AHP) in GIS platform. Figure 7.5 presents the seismic microzonation map of Delhi.

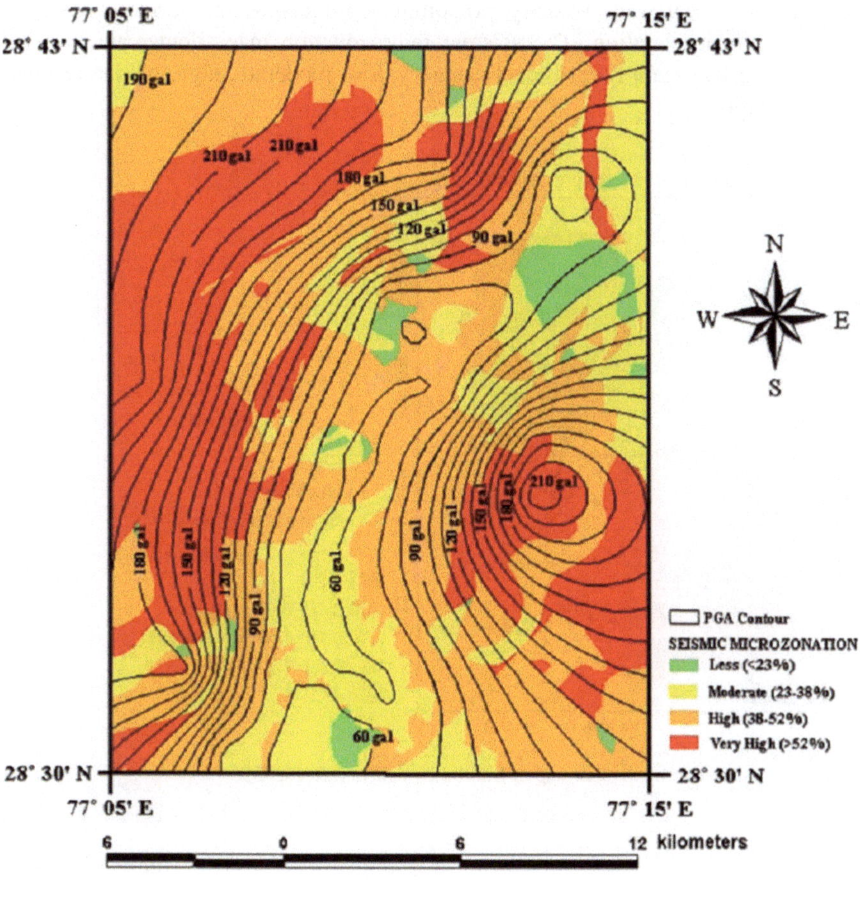

Fig. 7.5 Seismic microzonation map of Delhi (after [223])

7.4.3 Seismic Zonation of Sikkim Himalaya and Guwahati

Nath et al. [235] have presented the microzonation works carried for Sikkim Himalaya and Guwahati, which include multi-criteria seismic hazard analysis through thematic integration of influencing factors. Scenario earthquakes of M_W 8.3 and 8.7 were used for the seismic hazard analysis for two regions using dynamic finite fault simulation technique. Seismic hazard microzonation map of the Sikkim Himalaya (Fig. 7.6) was prepared after integration of geomorphological themes and seismological themes using the AHP in a GIS platform. Geomorphological theme for Sikkim Himalaya includes surface geology, soil cover, slope, rock outcrop, and landslide, while the seismological themes consist of peak ground acceleration and predominant frequency. Similarly seismic microzonation map of Guwahati (Fig. 7.7) was also prepared by integrating eight themes-geological and geomorphological, basement or bedrock, landuse, landslide, factor of safety for soil stability, shear wave velocity, predominant frequency, and surface consistent peak ground acceleration. The seismic microzonation maps demarcate the two regions into five main hazard level category, low, moderate, high, moderate high, and very high.

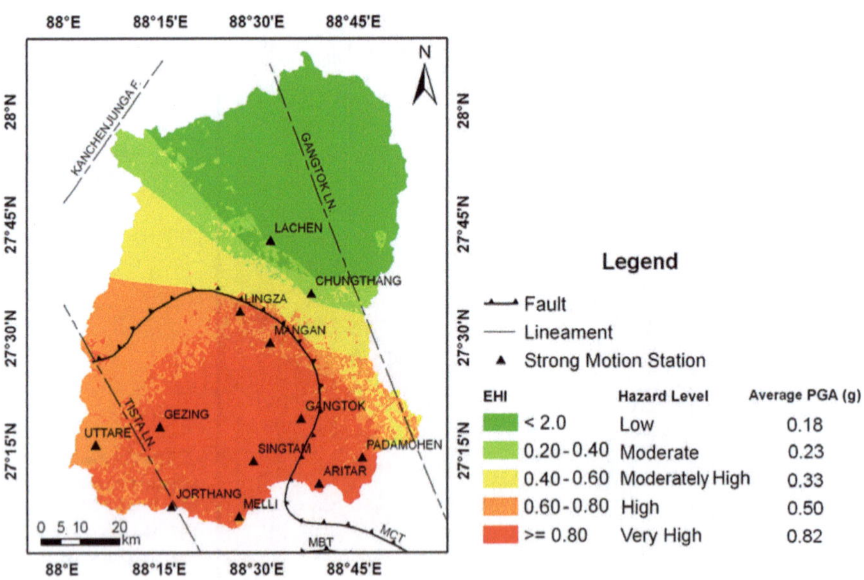

Fig. 7.6 Seismic microzonation map of Sikkim Himalaya (after [235])

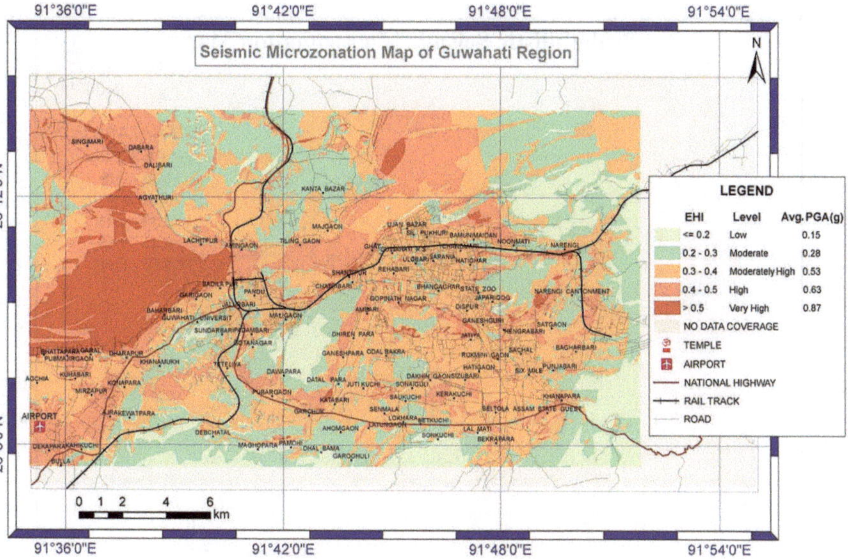

Fig. 7.7 Seismic microzonation map of Guwahati (after [235])

7.4.4 Seismic Zonation of Bangalore

Sitharam and Anbazhagan [328] have carried exhaustive studies for the seismic microzonation of Bangalore city. Seismotectonic map, comprising of earthquake events and seismic source was developed covering 350 km radius around Bangalore. Seismic hazard analysis for the study area has been carried out using deterministic as well as probabilistic approaches. Seismic site characterization for the region was carried out using both geotechnical and geophysical tests. They have also developed three dimensional sub surface borehole model for study area using geographical information system (GIS).

The study was further categorized into various NEHRP site classes based on V_{s30} measured using multichannel analysis of surface wave (MASW) tests. 1D site response analyses were carried out using synthetic earthquake time history and SHAKE, to estimate site effects. Liquefaction hazard for the entire Bangalore was estimated based on SPT N value. A final hazard index map for Bangalore was developed using Analytic Hierarchy Process (AHP) on a GIS platform by integrating various hazard parameters. Major seismological attributes considered in their study are peak Ground Acceleration (PGA) (from deterministic and probabilistic approach), amplification factor, predominant frequency, liquefaction, while the geomorphologic attributes include geology and geomorphology, rock depth/soil thickness, soil type and average shear wave velocity, drainage pattern and elevation level. Figure 7.8 presents the deterministic seismic microzonation maps for Bangalore.

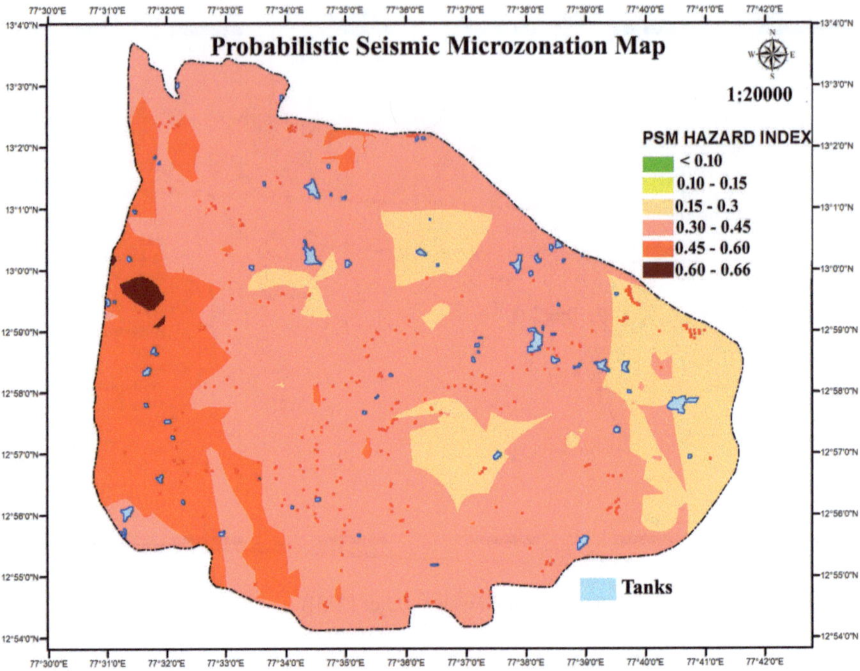

Fig. 7.8 Probabilistic seismic microzonation map of Bangalore (after [328])

7.5 Seismic Zonation Studies at Macro-Level

7.5.1 Seismic Zonation of Karnataka State

Similar to the micro-level zonation as described in the above section, a macro-level seismic zonation based on the hazard index value was also carried out for the entire state of Karnataka by James et al. [147]. Seismic macrozonation of a large area like a state or a country is an excellent tool for land-use planning. Macrozonation maps help the designers and planners to identify the sites seismically vulnerable and it can also provide inputs for site selection for critical structures nuclear/thermal power plants. These maps also provide the much needed information which can be functional in implementing appropriate mitigation works by the disaster management authority.

Similar to micro-level zonation work, the hazard index value was evaluated for the entire state of Karnataka by integrating a set of EHPs using AHP. Major EHPs considered in the evaluation of hazard index are (1) PHA at bedrock level, (2) the amplification factor, (3) SPT-N value required to prevent liquefaction at 3 m depth, and (4) the average shear wave velocity for top 30 m (V_{s30}). The highest weightage is given to PHA at bedrock level followed by amplification factor, SPT-N value

Table 7.4 Parameter and weights used for generating hazard index map at the macro-level (after [147])

Earthquake hazard parameter (EHP)	Weights
Peak horizontal acceleration (PHA)	4
Amplification factor (Amp.factor)	3
SPT-N value required to prevent liquefaction (Lq.Fn SPT)	2
Average shear wave velocity for top 30 m (V_{s30})	1

Table 7.5 Normalized weights for each (EHP) for macrozonation (after [151])

EHP	PHA	Amp. factor	Lq.Fn SPT	V_{s30}	Normalized weight
PHA	4/4	4/3	4/2	4/1	0.4
Amp.factor	3/4	3/3	3/2	3/1	0.3
Lq.Fn SPT	2/4	2/3	2/2	2/1	0.2
V_{s30}	1/4	1/3	1/2	1/1	0.1

required to prevent liquefaction and least weightage is given for (V_{s30}) as described in Table 7.4.

Once after assigning priorities to each of earthquake hazard parameters, a pairwise comparison matrix was constructed (Table 7.5). The each EHP was further sub-categorized into various ranges as in Table 7.6 and normalized ranks for each category was evaluated using Eq. (7.1). The hazard index for the entire state of Karnataka was evaluated using Eq. (7.4). Moreover the seismic landslide hazard, at a macro-level, is mainly restricted in Western Ghat sections of Karnataka, hence its effect is much concentrated in this region only.

$$\mathrm{HI} = \left(\frac{1}{\sum w}\right)\left[\begin{array}{c}\mathrm{PHA}_w \times \mathrm{PHA}_r + \mathrm{Amp.factor}_w \times \mathrm{Amp.factor}_r \\ +\mathrm{Lq.FnSPT}_w \times \mathrm{Lq.FnSPT}_w + V_{s30w} \times V_{s30r}\end{array}\right] \tag{7.4}$$

Figures 7.9 and 7.10 present the probabilistic seismic microzonation map for the state of Karnataka, for a return period of 475 years and 2500 years, respectively. Probabilistic macrozonation map for 475 year return period shows that 75–80% of Karnataka state has a hazard index value ranging from 0 to 0.4, hence falls in the category of low hazard region based on the classification proposed by Nath et al. [235]. Certain portions such as regions close to Bangalore in the south and Bidar in the north have the hazard index value ranging from 0.4 to 0.6 and hence can be categorized as regions having moderately high hazard.

Now referring to the probabilistic macrozonation map for 2500 year return period, the hazard index value for the interior regions of the Karnataka state varies from 0.2 to 0.6, falling in the category of low to moderately low hazard region. Even though the analysis shows that the Western Ghat regions are in low to moderate low hazard region, it should be kept in mind that this region has very high seismic landslide hazard. Similarly the hazard index for the regions close to Bangalore and

Table 7.6 Normalized rating for each EHP category for the macro-level hazard integration (after [150])

EHP	Value range	Weight	Rank	Normalized rank
PHA(g)	≤0.1	0.4	1	0
	0.10–0.20		2	0.25
	0.20–0.25		3	0.5
	0.25–0.30		4	0.75
	≥0.30		5	1
Amp. factor	1.0–2.0	0.3	1	0
	2.0–3.0		2	0.5
	≥3.0		3	1
Lq.Fn SPT	≤5	0.2	1	0
	5.0–15		2	0.33
	15–25		3	0.67
	≥25		4	1
V_{s30} (m/s)	≤100	0.1	6	1
	100–200		5	0.80
	200–300		4	0.60
	300–400		3	0.40
	400–500		2	0.20
	≥500		1	0

Bidar varies from 0.4 to 0.8 and can be classified as regions having moderately high to high hazard.

7.6 Guidelines for Seismic Zonation

The main objective of this book is to define the level of zonation to be adopted for various study areas in the country. As the seismic zonation program for various cities in India has been initiated by the government, the suitability of different levels of zonation for various regions within the country need to be studied. It is evident that the micro-level zonation requires very large resources in order to carry out geotechnical and geophysical testing and site effect estimation, hence it is not physically and financially viable to carry out micro-level seismic zonation for all these cities. The present study shows that the micro-level zonation is most suitable for critical facility like an NPP site situated even in a low seismic zone or a densely populated city or urban area situated in a high seismic zone. Nuclear power plants are very critical structures and any slightest of damages during an earthquake event will result in catastrophe. Similarly a densely populated city in a very high seismic zone escalates the risk to human lives and their property. Rigorous and extensive geotechnical and geophysical investigations are required for site characterization.

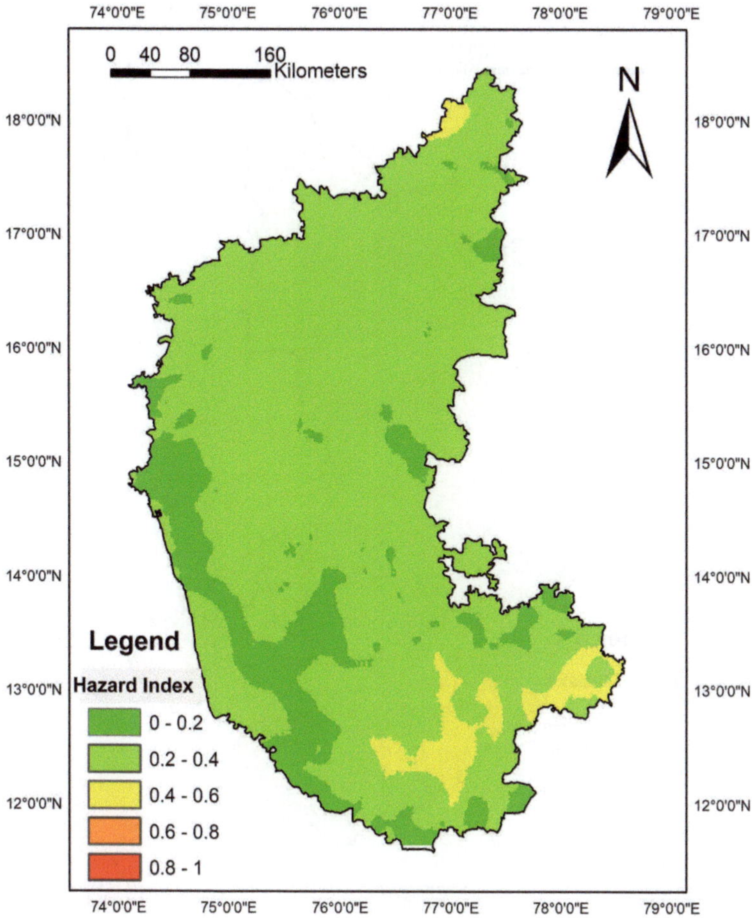

Fig. 7.9 Probabilistic seismic zonation map of Karnataka (475 year return period after [150])

Site effects should be estimated using numerical studies based on the geotechnical and geophysical test data. As the potential seismic risk is very high in the above two cases, the micro-level seismic zonation is very much required.

The meso-level zonations is preferred for the cities having population above 500,000, lying in the low to moderate seismic zones. The meso-level zonation is required as the level of seismic risk is high due to large population and infrastructure. The macro-level seismic zonation can very well be useful for analyzing the seismicity of that state or country and identifying vulnerable regions within it. This level of zonation can be adopted for cities as well, which are having very low population and infrastructures and are situated in very low seismic zones. As there are many methods available for site characterization and earthquake hazard estimation, the feasibility of various techniques for the regions at different scales

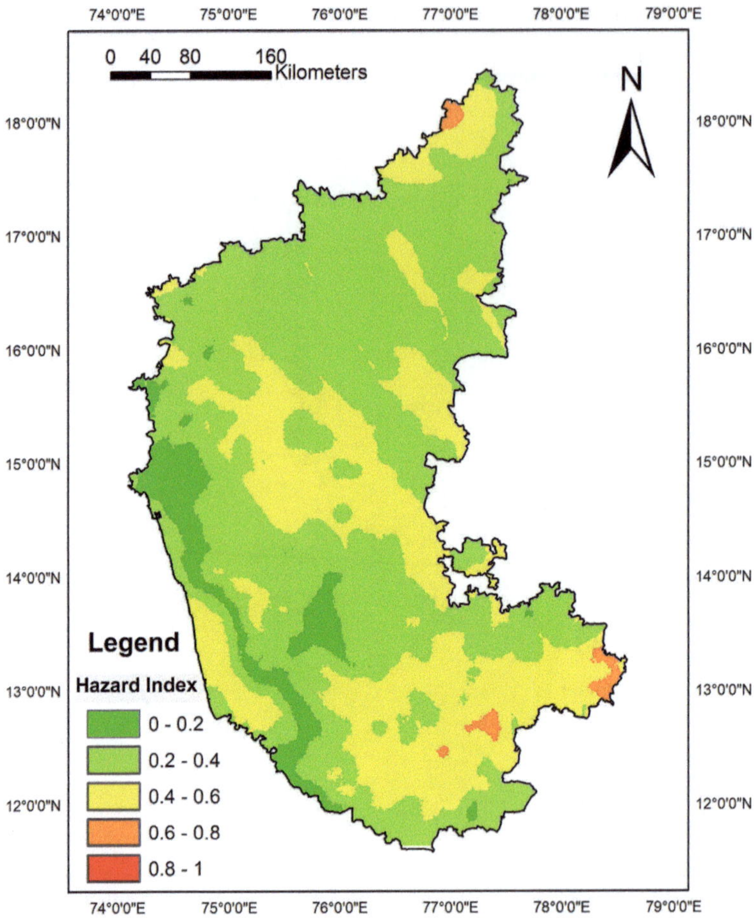

Fig. 7.10 Probabilistic seismic zonation map of Karnataka (2500 year return period after [151])

was reviewed in this study. Based on this, appropriate methods have been adopted for the micro and macro-level seismic zonation of the Kalpakkam NPP site and the Karnataka state, respectively. As the Kalpakkam NPP site area does not lie within the state of Karnataka, the results of the macro-level study were validated with the existing meso-level study carried out for the Bangalore city. The important recommendations of this book arrived after analyzing various factors are given below.

7.6.1 Delineation of Seismic Study Area

The seismic study area extending up to 300 km from the boundary of the study area should be identified and the details of the seismic events and the seismic sources need to be identified from this region.

7.6.2 Geological Consideration

Thorough regional and local geological studies including the basement configuration need to be done. Preparation of seismotectonic map considering all the active and paleo faults. The seismotectonic atlas published by GSI [316] may be taken as the base map. In addition to this the new lineaments or faults available in literature can also be considered.

7.6.3 Preparation of Earthquake Catalogue

Data need to be collected from various national and international agencies and from literature. It is recommended to decluster the data to remove the related events. The collected earthquake data may be in different magnitude scales. It has to be converted to the same scale in order to evaluate the seismicity parameters of the region. The earthquake data thus compiled need to be published in order to help compiling a national earthquake catalogue.

7.6.4 Attenuation Properties of the Region

Since the attenuation relations are not available for many regions of India, appropriate attenuation relations need to be selected. The relations developed for other parts of the world, which are having similar attenuation characteristics can be used. If the time history of regional earthquakes is available, then these attenuation relations need to be crosschecked with this data for validation.

7.6.5 Seismic Hazard Analysis

The PHA and spectral acceleration (S_a) at the bedrock level, for micro, meso, and macro-level can be evaluated using DSHA and PSHA methodologies. Since both methods are having its own advantages and disadvantages, it is better to evaluate the hazard using both the methods. Care should be taken in evaluating the maximum earthquake magnitude for each region and the recurrence rate. The initial estimates of ground motion parameters (PHA, PHV, or PHD) should be obtained at the bed rock level. The surface level values need to be calculated by considering the site response and local site conditions. Need to develop response spectra for bed rock conditions. It is advisable to get an estimate of ground motion duration also.

Seismic hazard analysis using multiple source/attenuation models along with logic tree methodology is found to be more effective for addressing epistemic uncertainties and providing realistic estimate of peak ground acceleration and spectral acceleration values.

7.6.6 Site Characterisation

Seismic site characterization based on geotechnical and geophysical tests is suitable for micro to meso-level zonation. However this methodology is not appropriate for a macro-level zonation as it is not economically and physically viable to carry out field tests for such a large area. Hence the site characterization based on topographic slope maps was found to be most suitable for the macro-level zonation. A comparison study has been made between the site characterization based on field tests and topographic slope maps at meso-level and a good correlation was found between them. Hence the study also shows that the site characterization based on topographic slope maps can be adopted for meso-level zonations also, however it is recommended for cities lying in low to moderate hazard zones.

7.6.7 Site Response Studies

Ground response analyses are very much required for micro-level zoning for estimating site amplification and surface level earthquake motion as it accurately captures the local soil effects. The empirical method of estimating site amplification based on site class coefficients and bedrock motion known as non-linear site amplification technique was found to be the most appropriate for macro-level zoning. For a meso-level zonation, the ground response analyses suitable for cities in high hazard zones having large population, and for the converse case, the non-linear site amplification techniques are preferable.

7.6.8 Liquefaction Hazard Assessment

The factor of safety against liquefaction can be evaluated based on the geotechnical data obtained from the filed. The liquefaction potential evaluation can be done

Table 7.7 Various methodologies and grid sizes suitable for the micro, meso and macro-level (after [151])

Earthquake hazards	Methodologies (grid size)		
	Micro-level	Meso-level	Macro-level
Seismic hazard at bedrock	Deterministic/ probabilistic approach (less than 1 km × 1 km)	Deterministic/ probabilistic approach (1 km × 1 km to 5 km × 5 km)	Deterministic/ probabilistic approach (5 km × 5 km to 10 km × 10 km)
Seismic site characterization	Geotechnical and geophysical testing (less than 1 km × 1 km)	Geotechnical and geophysical testing/ topographic slope map (1 km × 1 km to 5 km × 5 km)	Topographic slope map (5 km × 5 km to 10 km × 10 km)
Site amplification hazard	Ground response analyses (less than 1 km × 1 km)	Ground response analyses and Empirical method (1 km × 1 km to 5 km × 5 km)	Empirical method (5 km × 5 km to 10 km × 10 km)
Liquefaction hazard studies	Deterministic/ probabilistic method based on field test data (less than 1 km × 1 km)	Deterministic/ probabilistic method based on field test data (1 km × 1 km to 5 km × 5 km)	Performance based approach (Kramer and Mayfield [182]) (5 km × 5 km to 10 km × 10 km)

based on the deterministic or probabilistic methods. Liquefaction potential can be reported in terms of SPT and CPT values required to prevent liquefaction can also be developed (if required). Liquefaction factor of safety evaluated based on the available geotechnical data and using probabilistic techniques can then be used for the liquefaction hazard assessment. Same methodology can also be used to estimate liquefaction hazard at the meso-level for the cities lying in moderate to high hazard zone. For a macro-level study, the performance based approach for the evaluation of liquefaction hazard was found to be most appropriate and recommended.

Moreover, recommendations have also been made regarding suitability of various methodologies and the grid sizes to be adopted for site characterization and site effect studies at the micro, meso, and macro-level are presented in Table 7.7.

7.7 Summary

Seismic zonation is recognized as the most acceptable tool in seismic hazard assessment and risk evaluation. Making improvements on the conventional zonation maps and regional hazard maps, microzonation of a region generates detailed maps that predict the hazard at much smaller scales from one block of a city to another. Damage patterns of many recent earthquakes around the world, including the 1999 Chamoli and 2001 Bhuj earthquakes in India, have demonstrated that the soil conditions at a site may have a major effect on the ground shaking levels. Earthquake damage is commonly controlled by three interacting factors—source and path characteristics, local geological and geotechnical conditions, and design of the structures. Seismic zonation is the process of assessment of the first two factors to provide a basis for estimating and mapping potential damage to buildings, which in other words is quantification of the risk. Obviously, all of this would require analysis and presentation of a large amount of data consisting of geological, seismological, engineering specifications, etc. History of earthquakes, geometry and history of faults in the region, attenuation relationships, ground amplification, liquefaction susceptibility and landslide vulnerability are the other important inputs required.

Although several efforts are on for seismic zonation of selected regions in India, these studies are under development. While steps have been taken to strengthen research and development efforts on various aspects of earthquake related studies with the ultimate aim of mitigating the earthquake related disasters, efforts are also being made to prepare knowledge based products in the form of seismic hazard, vulnerability, and risk maps. The procedures are often modified or simplified and these are not adequately reflected in the final product. So also are the missing gaps in data, treatment of the data, etc. There is also an increasing tendency to do a partial or incomplete analysis of some available data, prepare some maps using such data, and refer to them as seismic zonation maps. Due to these inconsistencies in the procedures followed by various workers and the maps developed would be unacceptable to the end users.

In this book the suitability of different levels of seismic zonation for various regions with the country was assessed. The feasibility of various techniques for site characterization and estimation of site effects for regions at various scale levels were reviewed. Based on this, appropriate methods have been adopted and recommended for site characterization and estimation of local site effects for regions at different scale levels. Moreover, recommendations also have made regarding the grid size to be adopted for site characterization and site effect studies at micro, meso, and macro-level.

References

1. Adalier, K., Elgamal, A.: Mitigation of liquefaction and associated ground deformations by stone columns. Eng. Geol. **72**(3), 275–291 (2004)
2. AFPS, F.A.F.E.E.: Guidelines for Seismic Microzonation Studies. Delegation of Major Risks of the French Ministry of the Environment-Direction for Prevention, Pollution and Risks (1995)
3. Aki, K.: Space and time spectra of stationary stochastic waves, with special reference to microtremors. Bull. Earthq. Res. Inst., Univ. Tokyo **35**, 415–456 (1957)
4. Aki, K.: Maximum likelihood estimate of b in the formula logn= a-bm and its confidence limits. Bull. Earthq. Res. Inst. Tokyo **43**, 237–239 (1965)
5. Aki, K.: Generation and propagation of g waves from the Niigata earthquake of June 16, 1964: Part 1. A statistical analysis. Bull. Earthq. Res. Inst. Univ. Tokyo **44**(1), 23–72 (1966)
6. Aki, K.: Local site effects on ground motion. In: Earthquake Engineering and Soil Dynamics II-Recent Advances in Ground Motion Evaluation. Geotechnical Special Publication, vol. 20, pp. 103–155 (1988)
7. Alfaro, A., Pujades, L.G., Goula, X., Susagna, T., Navarro, M., Sanchez, J., Canas, J.: Preliminary map of soil's predominant periods in Barcelona using microtremors. In: Earthquake Microzoning, pp. 2499–2511. Springer, New York (2002)
8. Anbazhagan, P.: Site Characterization and Seismic Hazard Analysis with Local Site Effects for Microzonation of Bangalore. Indian Institute of Science, Bangalore (2007)
9. Anbazhagan, P., Sitharam, T.: Mapping of average shear wave velocity for Bangalore region: a case study. J. Environ. Eng. Geophys. **13**(2), 69–84 (2008)
10. Anbazhagan, P., Vinod, J., Sitharam, T.: Probabilistic seismic hazard analysis for Bangalore. Nat. Hazards **48**(2), 145–166 (2009)
11. Andersen, K.: Cyclic and static laboratory tests on Drammen clay. J. Geotech. Eng. Div. **106**(5), 499–529 (1980)
12. Andrews, D.: Objective determination of source parameters and similarity of earthquakes of different size. In: Earthquake Source Mechanics, pp. 259–267. American Geophysical Union, Washington (1986)
13. Andrus, R.D., Chung, R.M.: Ground Improvement Techniques for Liquefaction Remediation Near Existing Lifelines. US National Institute of Standards and Technology, Gaithersburg (1995)
14. Ansal, A., Erdik, M., Studer, J., Springman, S., Laue, J., Buchheister, J., Giardini, D., Faeh, D., Koksal, D.: Seismic microzonation for earthquake risk mitigation in turkey. In: 13th World Conference on Earthquake Engineering, Vancouver (2004)

T. G. Sitharam et al., *Comprehensive Seismic Zonation Schemes for Regions at Different Scales*, https://doi.org/10.1007/978-3-319-89659-5

15. Astroza, M., Monge, J.: Seismic microzones in the city of Santiago. Relation damage-geological unit. In: Proceedings of the Fourth International Conference on Seismic Zonation, vol. 3, pp. 25–29 (1991)
16. Atkinson, G., Boore, D.: Earthquake ground-motion prediction equations for eastern North America. Bull. Seismol. Soc. Am. **96**(6), 2181–2205 (2006)
17. Atkinson, G.M., Cassidy, J.F.: Integrated use of seismograph and strong-motion data to determine soil amplification: response of the Fraser river delta to the Duvall and Georgia Strait earthquakes. Bull. Seismol. Soc. Am. **90**(4), 1028–1040 (2000)
18. Bansal, B., Vandana, C.: Microzonation studies in India: DST initiatives. In: Proceedings of Workshop on Microzonation. Indian Institute of Science, Bangalore (2007)
19. Bard, P.Y., Tucker, B.E.: Underground and ridge site effects: a comparison of observation and theory. Bull. Seismol. Soc. Am. **75**(4), 905–922 (1985)
20. Bard, P.Y., Duval, A.M., Lebrun, B., Lachet, C., Riepl J. Hatzfeld, D.: Reliability of the H/V technique for site effects measurement: an experimental assessment.İ Seventh International Conference on Soil Dynamics and Earthquake Engineering, Istanbul (1997)
21. Basu, K.: A note on the Coimbatore earthquake of 8th February 1900. Indian J. Metamorph. Geol. **15**(2), 281–286 (1964)
22. Been, K., Jefferies, M.G.: A state parameter for sands. Géotechnique **35**(2), 99–112 (1985)
23. Bennington, N., Thurber, C., Roecker, S.: Three-dimensional seismic attenuation structure around the Safod site, Parkfield, California. Bull. Seismol. Soc. Am. **98**(6), 2934–2947 (2008)
24. Beresnev, I., Wen, K., Yeh, Y.: Nonlinear soil amplification: its corroboration in Taiwan. Bull. Seismol. Soc. Am. **85**(2), 496–515 (1995)
25. Bhatia, S., Kumar, M., Gupta, H.: A probabilistic seismic hazard map of India and adjoining regions. Ann. Geophys. **42**(6), 1153–1164 (1999)
26. Bian, H., Jia, Y., Shahrour, I.: A potential cost effective liquefaction mitigation countermeasure: induced partial saturation. In: AIP Conference Proceedings, vol. 1020, pp. 427–433. American Institute of Physics, New York (2008)
27. Bilham, R.: Earthquakes in India and the Himalaya: tectonics, geodesy and history. Ann. Geophys. **47**(2–3), 839–858 (2004)
28. BIS-1893: Indian Standard Criteria for Earthquake Resistant Design of Structures, Part 1 – Genera l Provisions and Buildings. Bureau of Indian Standards, New Delhi (2002)
29. BIS-2131: Method for Standard Penetration Test for Soils. Bureau of Indian Standards, New Delhi (2002)
30. Biswas, S., Gupta, A.D.: Some observations on the mechanism of earthquakes in the Himalaya and the Burmese arc. Tectonophysics **122**(3–4), 325–343 (1986)
31. Bommer, J., Scherbaum, F., Bungum, H., Cotton, F., Sabetta, F., Abrahamson, N.: On the use of logic trees for ground-motion prediction equations in seismic-hazard analysis. Bull. Seismol. Soc. Am. **95**(2), 377–389 (2005)
32. Bonilla, L.F., Steidl, J.H., Lindley, G.T., Tumarkin, A.G., Archuleta, R.J.: Site amplification in the San Fernando Valley, California: variability of site-effect estimation using the s-wave, coda, and h/v methods. Bull. Seismol. Soc. Am. **87**(3), 710–730 (1997)
33. Boominathan, A., Dodagoudar, G., Suganthi, A., Uma Maheswari, R.: Seismic hazard assessment of Chennai city considering local site effects. J. Earth Syst. Sci. **117**, 853–863 (2008)
34. Boore, D.: Stochastic simulation of high-frequency ground motions based on seismological models of the radiated spectra. Bull. Seismol. Soc. Am. **73**(6A), 1865–1894 (1983)
35. Boore, D.: Simulation of ground motion using the stochastic method. Pure Appl. Geophys. **160**, 635–676 (2003)
36. Boore, D.M.: Estimating vs (30)(or nehrp site classes) from shallow velocity models (depths < 30 m). Bull. Seismol. Soc. Am. **94**(2), 591–597 (2004)
37. Boore, D., Atkinson, G.: Ground-motion prediction equations for the average horizontal component of PGA, PGV, and 5%-damped PSA at spectral periods between 0.01 s and 10.0 s. Earthq. Spectra **24**(1), 99–138 (2008)

38. Boore, D., Joyner, W., Fumal, T.: Estimation of response spectra and peak accelerations from Western North American earthquakes: an Interim Report. US Geological Survey, Menlo Park (1993)
39. Borcherdt, R.: Estimates of site-dependent response spectra for design (methodology and justification). Earthq. Spectra **10**, 617–617 (1994)
40. Borcherdt, R.: Preliminary amplification estimates inferred from strong ground motion recordings of the Northridge earthquake of January 17, 1994. In: Proceeding of the International Workshop on Site Response Subjected to Strong Ground Motion (1996)
41. Borcherdt, R.D.: Empirical evidence for acceleration-dependent amplification factors. Bull. Seismol. Soc. Am. **92**(2), 761–782 (2002)
42. Borcherdt, R., Gibbs, J.: Effects of local geological conditions in the San Francisco bay region on ground motions and the intensities of the 1906 earthquake. Bull. Seismol. Soc. Am. **66**(2), 467–500 (1976)
43. Borcherdt, R.D., Glassmoyer, G.: Influences of local geology on strong and weak ground motions recorded in the San Francisco Bay region, p. 77. US Geological Survey Professional Paper (1994)
44. Borcherdt, R., Wentworth, C., Glassmoyer, G., Fumal, T., Mork, P., Gibbs, J.: On the observation, characterization and predictive GIS mapping of ground response in the San Francisco Bay region, California. In: Proceedings of 4th International Conference on Seismic Zonation, Stanford, pp. 545–552 (1991)
45. Bouckovalas, G., Kouretzis, G.: A review of soil and topography effects in Athens 09/07/199 (Greece) earthquake. In: Proceedings of Fourth International Conference on Recent Advances in Geotechnical Earthquake Engineering and Soil Dynamics, San Diego, March (in CDROM) (2001)
46. Bouckovalas, G.D., Papadimitriou, A.G.: Numerical evaluation of slope topography effects on seismic ground motion. Soil Dyn. Earthq. Eng. **25**(7), 547–558 (2005)
47. Boulanger, R.W., Hayden, R.F.: Aspects of compaction grouting of liquefiable soil. J. Geotech. Eng. **121**(12), 844–855 (1995)
48. Bowles, J.: Foundation Analysis and Design, 5th edn. McGraw-Hill, New Delhi (1988)
49. Brown, R.E.: Vibroflotation compaction of cohesionless soils. J. Geotech. Geoenviron. Eng. **103**, 1437–1451 (1977). ASCE 13415 Proceeding
50. Brune, J.N.: Preliminary results on topographic seismic amplification effect on a foam rubber model of the topography near Pacoima Dam. In: Proceeding of the 8th World Conference on Earthquake Engineering, vol. 2, pp. 663–670 (1984)
51. BSSC: NEHRP recommended provisions for seismic regulations for new buildings and other structures (FEMA 450), Part 1: Provisions. Building Seismic Safety Council for the Federal Emergency Management Agency, Washington (2003)
52. Budnitz, R.J., Apostolakis, G., Boore, D.M.: Recommendations for probabilistic seismic hazard analysis: guidance on uncertainty and use of experts. Tech. rep., Nuclear Regulatory Commission, Washington. Div. of Engineering Technology; Lawrence Livermore National Lab.; Electric Power Research Inst., Palo Alto; USDOE, Washington (1997)
53. Camp III, W.M., Camp, H.C., Andrus, R.D.: Liquefaction mitigation using air injection. In: Fifth International Conferences on Recent Advances in Geotechnical Earthquake Engineering and Soil Dynamics, San Diego (2010)
54. Campbell, K.: Empirical near-source attenuation relationships for horizontal and vertical components of peak ground acceleration, peak ground velocity, and pseudo-absolute acceleration response spectra. Seismol. Res. Lett. **68**(1), 154–179 (1997)
55. Campbell, K., Bozorgnia, Y.: Updated near-source ground-motion (attenuation) relations for the horizontal and vertical components of peak ground acceleration and acceleration response spectra. Bull. Seismol. Soc. Am. **93**(1), 314–331 (2003)
56. Campillo, M., Bard, P., Nicollin, F., Sánchez-Sesma, F.: The Mexico earthquake of September 19, 1985 – the incident wavefield in Mexico city during the great Michoacán earthquake and its interaction with the deep basin. Earthq. Spectra **4**(3), 591–608 (1988)

57. Casagrande, A.: Characteristics of cohesionless soils affecting the stability of slopes and earth fills. J. Boston Soc. Civ. Eng. 257–276 (1936). Reprinted in Contributions to soil mechanics, 1925–1940, Boston Society of Civil Engineers, Boston, Oct. 1940
58. Casagrande, A.: Liquefaction and Cyclic Deformation of Sands-A Critical Review. Harvard Soil Mechanics Series. Harvard University Press, Cambridge (1976)
59. Castro, G.: Liquefaction of sands. Harvard University Division of Engineering and Applied Mechanics (1969)
60. Castro, R.R., Condori, C., Romero, O., Jacques, C., Suter, M.: Seismic attenuation in northeastern Sonora, Mexico. Bull. Seismol. Soc. Am. **98**(2), 722–732 (2008)
61. Cetin, K., Seed, R., Der Kiureghian, A., Tokimatsu, K., Harder L. Jr., Kayen, R., Moss, R.: Standard penetration test-based probabilistic and deterministic assessment of seismic soil liquefaction potential. J. Geotech. Geoenviron. Eng. **130**(12), 1314–1340 (2004)
62. Chandra, U.: Earthquakes of peninsular India-a seismotectonic study. Bull. Seismol. Soc. Am. **67**(5), 1387–1413 (1977)
63. Chandra, U.: Seismicity, earthquake mechanisms and tectonics along the Himalayan mountain range and vicinity. Phys. Earth Planet. Inter. **16**(2), 109–131 (1978)
64. Chávez-García, F.J., Cuenca, J.: Site effects and microzonation in Acapulco. Earthq. Spectra **14**(1), 75–93 (1998)
65. Chávez-García, F.J., Sánchez, L., Hatzfeld, D.: Topographic site effects and HVSR a comparison between observations and theory. Bull. Seismol. Soc. Am. **86**(5), 1559–1573 (1996)
66. Cheng, X.H., Qiang, M., Zuan, Y., et al.: Dynamic response of liquefiable sand foundation improved by bio-grouting. Chin. J. Geotech. Eng. **35**(8), 1486–1495 (2013)
67. Chiou, B., Youngs, R.: PEER-NGA empirical ground motion model for the average horizontal component of peak acceleration and pseudo-spectral acceleration for spectral periods of 0.01 to 10 seconds. Interim Report Issued for USGS Review (2006)
68. Chiou, B.J., Youngs, R.R.: An nga model for the average horizontal component of peak ground motion and response spectra. Earthq. Spectra **24**(1), 173–215 (2008)
69. Chouet, B., De Luca, G., Milana, G., Dawson, P., Martini, M., Scarpa, R.: Shallow velocity structure of Stromboli Volcano, Italy, derived from small-aperture array measurements of Strombolian tremor. Bull. Seismol. Soc. Am. **88**(3), 653–666 (1998)
70. Chun, K.Y., Henderson, G.A.: Lg attenuation near the North Korean border with China, part II: model development from the 2006 nuclear explosion in North Korea. Bull. Seismol. Soc. Am. **99**(5), 3030–3038 (2009)
71. Commission on Geosciences, Environment and Resources (CGER): Ground Water at Yucca Mountain: How High Can It Rise? National Academies Press, Washington (1992)
72. Cornell, C.: Engineering seismic risk analysis. Bull. Seismol. Soc. Am. **58**(5), 1583–1606 (1968)
73. Costa, G., Panza, G., Suhadolc, P., Vaccari, F.: Zoning of the Italian territory in terms of expected peak ground acceleration derived from complete synthetic seismograms. J. Appl. Geochem. **30**(1), 149–160 (1993)
74. Cundall, P.A.: A computer model for simulating progressive large scale movements in blocky rock systems. In: Proceedings of ISRM Symposium, vol. 2, Nancy, pp. 129–136 (1971)
75. Davis, L.L., West, L.R.: Observed effects of topography on ground motion. Bull. Seismol. Soc. Am. **63**(1), 283–298 (1973)
76. DeJong, J.T., Fritzges, M.B., Nüsslein, K.: Microbially induced cementation to control sand response to undrained shear. J. Geotech. Geoenviron. Eng. **132**(11), 1381–1392 (2006)
77. Dinesh, S., Sitharam, T., Vinod, J.: Dynamic properties and liquefaction behaviour of granular materials using discrete element method. Curr. Sci. **87**, 1379–1387 (2004)
78. Dobry, R., Ladd, R.S., Yokel, F.Y., Chung, R.M., Powell, D.: Prediction of Pore Water Pressure Buildup and Liquefaction of Sands During Earthquakes by the Cyclic Strain Method, vol. 138. US Department of Commerce National Bureau of Standards, Washington, DC (1982)

79. Dunbar, P.K., Lockridge, P.A., Whiteside, L.S.: Catalog of Significant Earthquakes, 2150 BC-1991 AD: Including Quantitative Casualties and Damage. US Department of Commerce, National Oceanic and Atmospheric Administration, Environmental Data and Information Service, National Geophysical Data Center, Washington (1992)
80. Duputel, Z., Rivera, L., Kanamori, H., Hayes, G.: W phase source inversion for moderate to large earthquakes (1990–2010). Geophys. J. Int. **189**(2), 1125–1147 (2012)
81. El Mohtar, C.S.: Pore fluid engineering: an autoadaptive design for liquefaction mitigation. Doctoral dissertation, Purdue University (2008)
82. El Mohtar, C., Bobet, A., Santagata, M., Drnevich, V., Johnston, C.: Liquefaction mitigation using bentonite suspensions. J. Geotech. Geoenviron. Eng. **139**(8), 1369–1380 (2012)
83. El-Sayed, A., Vaccari, F., Panza, G.: Deterministic seismic hazard in Egypt. Geophys. J. Int. **144**(3), 555–567 (2001)
84. El Shamy, U., Abdelhamid, Y.: Modeling granular soils liquefaction using coupled lattice Boltzmann method and discrete element method. Soil Dyn. Earthq. Eng. **67**, 119–132 (2014)
85. Eseller-Bayat, E., Yegian, M., Alshawabkeh, A., Gokyer, S.: Prevention of liquefaction during earthquakes through induced partial saturation in sands. In: Geotechnical Engineering: New Horizons, pp. 188–194. IOS Press, Amsterdam (2012)
86. Eurocode-8: BS-EN 1998-1, Design of structures for earthquake resistance? part 1: general rules, seismic actions and rules for buildings. European Committee for Standardization, Brussels (2003)
87. Evernden, J., Thomson, J.: Predicting seismic intensities. In: Evaluating Earthquake Hazards in the Los Angeles Region – An Earth-Science Perspective, vol. 1360, pp. 151–202. US Geological Survey Professional Paper (1985)
88. Faccioli, E.: Seismic amplification in the presence of geological and topographic irregularities. In: Proceedings of the Second International Conference on Recent Advances in Geotechnical Earthquake Engineering and Soil Dynamics, pp. 11–15 (1991)
89. Faccioli, E., Pessina, V.: The Catania Project: Earthquake Damage Scenarios for High Risk Area in the Mediterranean. CNR-Gruppo Nazionale per la Difesa dai Terremoti, Roma (2000)
90. Fah, D., Iodice, C., Suhadolc, P., Panza, G.: Application of numerical simulations for a tentative seismic microzonation of the city of Rome. Ann. Geophys. **38**(5–6), 607–615 (1995)
91. Fäh, D., Kind, F., Lang, K., Giardini, D.: Earthquake scenarios for the city of Basel. Soil Dyn. Earthq. Eng. **21**(5), 405–413 (2001)
92. Field, E., Jacob, K.: The theoretical response of sedimentary layers to ambient seismic noise. Geophys. Res. Lett. **20**(24), 2925–2928 (1993)
93. Finn, W., Pickering, D., Bransby, P.: Sand liquefaction in triaxial and simple shear tests. J. Soil Mech. Found. Div. **97**(SM4), 639–659 (1971)
94. Ford, S.R., Dreger, D.S., Mayeda, K., Walter, W.R., Malagnini, L., Phillips, W.S.: Regional attenuation in northern California: a comparison of five 1d q methods. Bull. Seismol. Soc. Am. **98**(4), 2033–2046 (2008)
95. Foti, S.: Multistation methods for geotechnical characterization using surface waves. Ph.D. thesis, Politecnico di Torino (2000)
96. Frankel, A.: Mapping seismic hazard in the central and eastern United States. Seismol. Res. Lett. **66**(4), 8–21 (1995)
97. Fumal, T., Tinsley, J.: Mapping shear-wave velocities of near-surface geologic materials. In: Evaluating Earthquake Hazards in the Los Angeles Region: An Earth-Science Perspective, vol. 1360, pp. 101–126. US Geological Survey Professional Paper (1985)
98. Gallagher, P.M.: Passive site remediation for mitigation of liquefaction risk. Ph.D. thesis (2000)
99. Gallagher, P.M., Mitchell, J.K.: Influence of colloidal silica grout on liquefaction potential and cyclic undrained behavior of loose sand. Soil Dyn. Earthq. Eng. **22**(9), 1017–1026 (2002)
100. Gallagher, P.M., Pamuk, A., Abdoun, T.: Stabilization of liquefiable soils using colloidal silica grout. J. Mater. Civ. Eng. **19**(1), 33–40 (2007)
101. Ganesha Raj, K., Nijagunappa, R.: Major lineaments of Karnataka state and their relation to seismicity-remote sensing based analysis. J. Geol. Soc. India **63**, 430–439 (2004)

102. Gangrade, B., Arora, S.: Seismicity of the Indian peninsular shield from regional earthquake data. Pure Appl. Geophys. **157**(10), 1683–1705 (2000)
103. Gazetas, G.: Seismic response of earth dams: some recent developments. Soil Dyn. Earthq. Eng. **6**(1), 2–47 (1987)
104. Gazetas, G., Dakoulas, P.: Seismic analysis and design of rockfill dams: state-of-the-art. Soil Dyn. Earthq. Eng. **11**(1), 27–61 (1992)
105. Geli, L., Bard, P., Jullien, B.: The effect of topography on earthquake ground motion: a review and new results. Bull. Seismol. Soc. Am. **78**(1), 42–63 (1988)
106. Ghasemi, H., Zare, M., Fukushima, Y., Koketsu, K.: An empirical spectral ground-motion model for Iran. J. Seismol. **13**(4), 499–515 (2009)
107. Giardini, D.: Systematic analysis of deep seismicity: 200 centroid-moment tensor solutions for earthquakes between 1977 and 1980. Geophys. J. Int. **77**(3), 883–914 (1984)
108. Gitterman, Y., Zaslavsky, Y., Shapira, A., Shtivelman, V.: Empirical site response evaluations: case studies in Israel. Soil Dyn. Earthq. Eng. **15**(7), 447–463 (1996)
109. Guha, S., Basu, P.: Catalogue of earthquakes (=> m 3.0) in peninsular India. Tech. rep., Atomic Energy Regulatory Board (1993)
110. Gupta, I.: Response spectral attenuation relations for in-slab earthquakes in Indo-Burmese subduction zone. Soil Dyn. Earthq. Eng. **30**(5), 368–377 (2010)
111. Gutenberg, B., Richter, C.: Magnitude and energy of earthquakes. Ann. Geophys. **83**(2147), 183–185 (1936)
112. Gutenberg, B., Richter, C.: Frequency of earthquakes in California. Bull. Seismol. Soc. Am. **34**(4), 185–188 (1944)
113. Gutenberg, B.U., Richter, C.F.: Seismicity of the Earth and Related Phenomena. Princeton University Press, Princeton (1954)
114. Gutenberg, B., Richter, C.: Magnitude and energy of earthquakes. Ann. Geophys. **9**, 1–15 (1956)
115. Haghshenas, E., Bard, P., Theodulidis, N.: Empirical evaluation of microtremor h/v spectral ratio. Bull. Earthq. Eng. **6**(1), 75–108 (2008)
116. Haldavnekar, V., Bobet, A., Santagata, M., Drnevich, V.: Soil treatment with a thixotropic fluid: an autoadaptive design for liquefaction prevention. In: Proceedings of the 11th International Conference on Soil Dynamics and Earthquake Engineering and 3rd International Conference on Earthquake Geotechnical Engineering, vol. 2, pp. 553–560 (2003)
117. Hall, L., Bodare, A.: Analyses of the cross-hole method for determining shear wave velocities and damping ratios. Soil Dyn. Earthq. Eng. **20**(1), 167–175 (2000)
118. Hanzawa, H., Itoh, Y., Suzuki, K.: Shear characteristics of a quick sand in the Arabian gulf. Soils Found. **19**(4), 1–15 (1979)
119. Harada, N., Towhata, I., Takatsu, T., Tsunoda, S., Sesov, V.: Development of new drain method for protection of existing pile foundations from liquefaction effects. Soil Dyn. Earthq. Eng. **26**(2), 297–312 (2006)
120. Hartzell, S.: Earthquake aftershocks as Green's functions. Geophys. Res. Lett. **5**(1), 1–4 (1978)
121. Hartzell, S., Cranswick, E., Frankel, A., Carver, D., Meremonte, M.: Variability of site response in the Los Angeles urban area. Bull. Seismol. Soc. Am. **87**(6), 1377–1400 (1997)
122. Hartzell, S., Carver, D., Cranswick, E., Frankel, A.: Variability of site response in Seattle, Washington. Bull. Seismol. Soc. Am. **90**(5), 1237–1250 (2000)
123. Hashash, Y.M.A., Groholski, D., Phillips, C., Park, D., Musgrove, M.: DEEPSOIL 4.0, User Manual and Tutorial. US Geological Survey, Washington (2011)
124. Hayashi, K., Suzuki, H.: Cmp cross-correlation analysis of multi-channel surface-wave data. Explor. Geophys. **35**(1), 7–13 (2004)
125. He, J., Ivanov, V., Chu, J.: Mitigation of liquefaction of saturated sand using biogas. Géotechnique **63**, 267–275 (2013)
126. Heaton, T., Tajima, F., Mori, A.: Estimating ground motions using recorded accelerograms. Surv. Geophys. **8**(1), 25–83 (1986)

127. Howell, R., Rathje, E.M., Kamai, R., Boulanger, R.: Centrifuge modeling of prefabricated vertical drains for liquefaction remediation. J. Geotech. Geoenviron. Eng. **138**(3), 262–271 (2012)
128. Huang, Y., Wen, Z.: Recent developments of soil improvement methods for seismic liquefaction mitigation. Nat. Hazards **76**(3), 1927–1938 (2015)
129. Iai, S., Koizumi, K., Noda, S., Tsuchida, H.: Large Scale Model Tests and Analyses of Gravel Drains. Un'yushō Kōwan Gijutsu Kenkyūjo, Yokosuka (1988)
130. IBC: International Building Code. International Code Council, Washington (2009)
131. Idriss, I., Boulanger, R.: Semi-empirical procedures for evaluating liquefaction potential during earthquakes. Soil Dyn. Earthq. Eng. **26**(2), 115–130 (2006)
132. Idriss, I., Sun, J.: User's Manual for shake91. Center for Geotechnical Modeling, Department of Civil Engineering, University of California, Davis (1992)
133. Imai, T.: P and s wave velocities of the ground in Japan. In: Proceeding of the 9th ICSMFE, vol. 2, pp. 257–260 (1977)
134. Imai, T.: Correlation of n-value with s-wave velocity and shear modulus. In: Proceedings of the Second European Symposium on Penetration Testing, pp. 67–72 (1981)
135. Imai, T., Yoshimura, Y.: Elastic wave velocity and soil properties in soft soil. Tsuchito-Kiso **18**(1), 17–22 (1970)
136. Imai, T., Yoshimura, M.: The relation of mechanical properties of soils to p and s wave velocities for soil ground in Japan. Urana Research Institute, OYO Corp., Tokyo (1972)
137. Irikura, K.: Semi-empirical estimation of strong ground motions during large earthquakes. Bull. Disaster Prev. Res. Inst. Kyoto Univ. **33**(2), 63–104 (1983)
138. Ishihara, K.: Stability of natural deposits during earthquakes. In: Proceedings of the 11th International Conference on Soil Mechanics and Foundation Engineering, vol. 1, pp. 321–376 (1985)
139. Ishihara, K.: Dynamic properties of soils and gravels from laboratory tests. In: Soil Dynamics and Geotechnical Engineering, pp. 1–17. Balkema, Rotterdam (1993)
140. Ishihara, K.: Soil Behaviour in Earthquake Geotechnics. Clarendon Press/Oxford University Press, Oxford (1996)
141. Ishihara, K.: Liquefaction resistance of nearly saturated sand as correlated with longitudinal wave velocity. In: Poromechanics, pp. 583–586. Balkema, Rotterdam (1998)
142. Ishihara, K., Li, S.: Liquefaction of saturated sand in triaxial torsion shear test. Soils Found. **12**(2), 19–39 (1972)
143. Ishihara, K., Tatsuoka, F., Yasuda, S.: Undrained deformation and liquefaction of sand under cyclic stresses. Soils Found. **15**(1), 29–44 (1975)
144. ISSMGE, T.: Manual for Zonation on Seismic Geotechnical Hazard. International Society of Soil Mechanics and Geotechnical Engineering (ISSMGE). The Japanese Geotechnical Society, Tokyo (1999)
145. Iwasaki, T., Tokida, K., Tatsuoka, F., Watanabe, S., Yasuda, S., Sato, H.: Microzonation for soil liquefaction potential using simplified methods. In: Proceedings 3rd International Conference on Microzonation, pp. 1319–1330 (1982)
146. Iyengar, R., Ghosh, S.: Microzonation of earthquake hazard in greater Delhi area. Curr. Sci. **87**(9), 1193–1202 (2004)
147. Iyisan, R.: Correlations between shear wave velocity and in-situ penetration test results. Tech. J. Chamber Civil Eng. Turkey **7**, 371–374 (1996)
148. Jafari, M., Asghari, A., Rahmani, I.: Empirical correlation between shear wave velocity (vs) and spt-N value for south of Tehran soils. In: Proceedings of the 4th International Conference on Civil Engineering, Tehran (1997)
149. Jaiswal, K., Sinha, R.: Probabilistic seismic-hazard estimation for peninsular India. Bull. Seismol. Soc. Am. **97**(1B), 318–330 (2007)
150. James, N.: Site characterization and assessment of various earthquake hazards for micro and macro-level seismic zonations of regions in the peninsular India. Ph.D. thesis, Indian Institute of Science, Bangalore (2013)

151. James, N., Sitharam, T.: Seismic zonations at micro and macro-level for regions in the peninsular India. Int. J. Geotech. Earthq. Eng. **7**(2), 35–63 (2016)
152. James, N., Sitharam, T., Vipin, K.: Assessment of liquefaction potential index using deterministic and probabilistic approaches. Int. J. Geotech. Earthq. Eng. **3**(2), 60–76 (2012)
153. James, N., Sitharam, T., Padmanabhan, G., Pillai, C.: Seismic microzonation of a nuclear power plant site with detailed geotechnical, geophysical and site effect studies. Nat. Hazards **71**(1), 419–462 (2014)
154. Johnson, J.M., Tanioka, Y., Ruff, L.J., Satake, K., Kanamori, H., Sykes, L.R.: The 1957 great Aleutian earthquake. In: Shallow Subduction Zones: Seismicity, Mechanics and Seismic Potential, pp. 3–28. Springer, Berlin (1994)
155. Johnston, A.: Seismic moment assessment of earthquakes in stable continental regions, instrumental seismicity. Geophys. J. Int. **124**(2), 381–414 (1996)
156. Joyner, W., Boore, D.: Peak horizontal acceleration and velocity from strong-motion records including records from the 1979 imperial valley, California, earthquake. Bull. Seismol. Soc. Am. **71**(6), 2011–2038 (1981)
157. Joyner, W., Fumal, T.: Use of measured shear-wave velocity for predicting geologic site effects on strong ground motion. In: Proceeding of the 8th World Conference on Earthquake Engineering, vol. 2, pp. 777–783 (1984)
158. Kagami, H., Okada, S., Ohta, Y.: Versatile application of dense and precision seismic intensity data by an advanced questionnaire survey. In: Proceeding Ninth World Conference on Earthquake Engineering, vol. 8, pp. 937–942 (1988)
159. Kaila, K., Sarkar, D.: Atlas of Isoseismal Maps of Major Earthquakes in India. National Geophysical Research Institute, Hyderabad (1978)
160. Kalkan, E., Gülkan, P., Yilmaz, N., Çelebi, M.: Reassessment of probabilistic seismic hazard in the Marmara region. Bull. Seismol. Soc. Am. **99**(4), 2127–2146 (2009)
161. Kanai K., Osada, T., Tanaka, T.: An investigation into the nature of microtremors. Bull. Earthq. Res. Inst. **32**, 199–209 (1954)
162. Kanai, K.: On microtremors VIII. Bull. Earthq. Res. Inst. **39**, 97–114 (1961)
163. Kanai, K., Tanaka, T., Morishita, T., Osada, K.: Observation of microtremors. xi.: Matsushiro earthquake swarm area. Bull. Earthq. Res. Inst. Tokyo Univ. **44**(3), 1297–1333 (1967). http://hdl.handle.net/2261/12296
164. Kanamori, H.: Seismological evidence for a lithospheric normal faulting – the Sanriku earthquake of 1933. Phys. Earth Planet. Inter. **4**(4), 289–300 (1971)
165. Kanamori, H.: Re-examination of the earth's free oscillations excited by the Kamchatka earthquake of November 4, 1952. Phys. Earth Planet. Inter. **11**(3), 216–226 (1976)
166. Kanamori, H.: The energy release in great earthquakes. J. Geophys. Res. **82**(20), 2981–2987 (1977)
167. Kanamori, H., Anderson, D.L.: Theoretical basis of some empirical relations in seismology. Bull. Seismol. Soc. Am. **65**(5), 1073–1095 (1975)
168. Karol, R.H.: Chemical grouts and their properties. In: Grouting in Geotechnical Engineering, pp. 359–377. American Society of Civil Engineers, New York (1982)
169. Kaul, M.: Stochastic characterization of earthquakes through their response spectrum. Earthq. Eng. Struct. Dyn. **6**(5), 497–509 (1978)
170. Kelkar, Y.: Earthquakes experienced in Maharashtra during the last 300 years. Daily Kesari Poona (1968)
171. Khattri, K., Rogers, A., Perkins, D., Algermissen, S.: A seismic hazard map of India and adjacent areas. Tectonophysics **108**(1), 93–134 (1984)
172. Khoubbi-Al, I., Adams, J.: Local site effects in Ottawa, Canada–first results from a strong motion network. In: Proceedings of the 13th World Conference on Earthquake Engineering, Paper, 2504 (2004)
173. Kiratzi, A., Karakaisis, G., Papadimitriou, E., Papazachos, B.: Seismic source-parameter relations for earthquakes in Greece. Pure Appl. Geophys. **123**(1), 27–41 (1985)
174. Kokusho, T.: Cyclic triaxial test of dynamic soil properties for wide strain range. Soils Found. **20**(2), 45–60 (1980)

175. Kolathayar, S., Sitharam, T.: Characterization of regional seismic source zones in and around India. Seismol. Res. Lett. **83**(1), 77–85 (2012)
176. Kolathayar, S., Sitharam, T.: Comprehensive probabilistic seismic hazard analysis of the Andaman–Nicobar regions. Bull. Seismol. Soc. Am. **102**(5), 2063–2076 (2012)
177. Kolathayar, S., Sitharam, T., Vipin, K.: Deterministic seismic hazard macrozonation of India. J. Earth Syst. Sci. **121**(5), 1351–1364 (2012)
178. Kolathayar, S., Vipin, K., Sitharam, T.: Recent seismicity in India and adjoining regions. Int. J. Earth Sci. Eng. **5**(1), 51–59 (2012)
179. Konno, K., Ohmachi, T.: Ground-motion characteristics estimated from spectral ratio between horizontal and vertical components of microtremor. Bull. Seismol. Soc. Am. **88**(1), 228–241 (1998)
180. Koulakov, I., Bindi, D., Parolai, S., Grosser, H., Milkereit, C.: Distribution of seismic velocities and attenuation in the crust beneath the North Anatolian Fault (Turkey) from local earthquake tomography. Bull. Seismol. Soc. Am. **100**(1), 207–224 (2010)
181. Kramer, S.: Geotechnical Earthquake Engineering. Pearson Education, Delhi (1996). Reprinted 2003
182. Kramer, S., Mayfield, R.: Return period of soil liquefaction. J. Geotech. Geoenviron. Eng. **133**(7), 802–813 (2007)
183. Krishnaswamy, N., Isaac, N.T.: Liquefaction potential of reinforced sand. Geotext. Geomembr. **13**(1), 23–41 (1994)
184. Kumar, A.: Software for generation of spectrum compatible time history. In: 13th World Conference on Earthquake Engineering Vancouver (2004)
185. Kumar, P., Yuan, X., Kumar, M.R., Kind, R., Li, X., Chadha, R.: The rapid drift of the Indian tectonic plate. Nature **449**(7164), 894–897 (2007)
186. Kumar, A., Anbazhagan, P., Sitharam, T.: Seismic hazard analysis of Lucknow considering local and active seismic gaps. Nat. Hazards **69**(1), 327–350 (2013)
187. Lachet, C., Hatzfeld, D., Bard, P.Y., Theodulidis, N., Papaioannou, C., Savvaidis, A.: Site effects and microzonation in the city of Thessaloniki (Greece) comparison of different approaches. Bull. Seismol. Soc. Am. **86**(6), 1692–1703 (1996)
188. Langston, C.A.: Structure under mount rainier, Washington, inferred from teleseismic body waves. J. Geophys. Res. Solid Earth **84**(B9), 4749–4762 (1979)
189. Lapajne, J., Motnikar, B., Zupančič, P.: Probabilistic seismic hazard assessment methodology for distributed seismicity. Bull. Seismol. Soc. Am. **93**(6), 2502–2515 (2003)
190. Lee, M.K., Finn, W.: DESRA-2: dynamic effective stress response analysis of soil deposits with energy transmitting boundary including assessment of liquefaction potential. Department of Civil Engineering, University of British Columbia (1978)
191. Lee, S.J., Chen, H.W., Liu, Q., Komatitsch, D., Huang, B.S., Tromp, J.: Three-dimensional simulations of seismic-wave propagation in the Taipei basin with realistic topography based upon the spectral-element method. Bull. Seismol. Soc. Am. **98**(1), 253–264 (2008)
192. Lermo, J., Chávez-García, F.J.: Site effect evaluation using spectral ratios with only one station. Bull. Seismol. Soc. Am. **83**(5), 1574–1594 (1993)
193. Liao, S., Whitman, R.: A Catalog of Liquefaction and Non-liquefaction Occurrences During Earthquakes. Department of Civil Engineering, MIT, Cambridge (1986)
194. Liao, S., Veneziano, D., Whitman, R.: Regression models for evaluating liquefaction probability. J. Geotech. Eng. **114**(4), 389–411 (1988)
195. Lin, C.P., Chang, C.C., Chang, T.S.: The use of masw method in the assessment of soil liquefaction potential. Soil Dyn. Earthq. Eng. **24**(9), 689–698 (2004)
196. Liu, R.F., Chen, Y.T., Ren, X., Xu, Z.G., Sun, L., Yang, H., Liang, J.H., Ren, K.X.: Comparison between different earthquake magnitudes determined by china seismograph network. Acta Seismol. Sin. **20**, 497–506 (2007)
197. Loh, C.H., Hwang, J.Y., Shin, T.C.: Observed variation of earthquake motion across a Basin – Taipei city. Earthq. Spectra **14**(1), 115–133 (1998)
198. López, A.M., Okal, E.A.: A seismological reassessment of the source of the 1946 aleutian 'tsunami' earthquake. Geophys. J. Int. **165**(3), 835–849 (2006)

199. Lungu, D., Aldea, A., Cornea, T., Arion, C.: Seismic microzonation of the city of Bucharest. In: 6th International Conference on Seismic Zonation, Palm Springs (2000)
200. Lunne, T., Robertson, P., Powell, J.: Cone Penetration Testing. Geotechnical Practice (1997)
201. Ma, S., Archuleta, R.J., Page, M.T.: Effects of large-scale surface topography on ground motions, as demonstrated by a study of the San Gabriel mountains, Los Angeles, California. Bull. Seismol. Soc. Am. **97**(6), 2066–2079 (2007)
202. MacMurdo, J.: Papers relating to the earthquake which occurred in India in 1819. Philos. Mag. **63**, 105–177 (1824)
203. Maher, M.H., Woods, R.D.: Dynamic response of sand reinforced with randomly distributed fibers. J. Geotech. Eng. **116**(7), 1116–1131 (1990)
204. Malagnini, L., Rovelli, A., Hough, S., Seeber, L.: Site amplification estimates in the Garigliano Valley, Central Italy, based on dense array measurements of ambient noise. Bull. Seismol. Soc. Am. **83**(6), 1744–1755 (1993)
205. Malagnini, L., Mayeda, K., Uhrhammer, R., Akinci, A., Herrmann, R.B.: A regional ground-motion excitation/attenuation model for the San Francisco region. Bull. Seismol. Soc. Am. **97**(3), 843–862 (2007)
206. Mallet, R.: Great Neapolitan Earthquake of 1857. Chapman and Hall, London (1862)
207. Mancuso, C.: Damping of soil by cross hole method. In: Proceedings of the International Conference on Soil Mechanics and Foundation Engineering-International Society for Soil Mechanics and Foundation Engineering, vol. 3, pp. 1337–1337. AA Balkema, Rotterdam (1994)
208. Marcellini, A., Daminelli, R., Pagani, M., Riva, F., Tento, A., Crespellani, T., Madiai, C., Vannucchi, G., Frassineti, G., Martelli, L., et al.: Seismic microzonation of some municipalities of the Rubicone area (Emilia-Romagna region). In: Bisch, P., Labb, P., Pecker, A. (eds.) Proceedings of the Eleventh European Conference on Earthquake Engineering, Invited Lectures Volume, pp. 339–350. Balkema, Rotterdam (1999)
209. Marcuson, W., Hynes, M., Franklin, A.: Evaluation and use of residual strength in seismic safety analysis of embankments. Earthq. Spectra **6**(3), 529–572 (1990)
210. Mashiri, M., Vinod, J., Sheikh, M.N.: Liquefaction potential and dynamic properties of sand-tyre chip (STCh) mixtures. Geotech. Test. J. **39**, 69–79 (2016)
211. Matsuoka, M., Wakamatsu, K., Fujimoto, K., Midorikawa, S.: Nationwide site amplification zoning using GIS-based Japan engineering geomorphologic classification map. In: Proceedings of the 9th International Conference on Structural Safety and Reliability, pp. 239–246 (2005)
212. Mayer-Rosa, D., Jimenez, M.: Seismic zoning, recommendations for Switzerland. Landeshydrologie und-geologie, geologischer bericht (1999)
213. Mayne, P.W., Rix, G.J.: Correlations between shear wave velocity and cone tip resistance in natural clays. Soils Found. **35**(2), 107–110 (1995)
214. Medvedev, J.: Engineering Seismology, p. 260. Academia Nauk Press, Moscow (1962)
215. Menon, A., Ornthammarath, T., Corigliano, M., Lai, C.: Probabilistic seismic hazard macrozonation of Tamil Nadu in southern India. Bull. Seismol. Soc. Am. **100**(3), 1320–1341 (2010)
216. Midorikawa, S.: Prediction of isoseismal map in the Kanto plain due to hypothetical earthquake. J. Struct. Eng. B **33**, 43–48 (1987)
217. Millea, M.: Liquefaction mitigation technology. Tech. rep., Naval Civil Engineering Lab Port Hueneme (1990)
218. Mitrani, H., Madabhushi, S.: Centrifuge modelling of inclined micro-piles for liquefaction remediation of existing buildings. Geomech. Geoeng.: Int. J. **3**(4), 245–256 (2008)
219. Mitsunan, T., Isemoto, N., Yasui, M., Shimoda, M., Bessho, M., Hanada, M., et al.: A use of super lime pile method for improvement of saturated loose sandy soils. In: The Third International Offshore and Polar Engineering Conference International Society of Offshore and Polar Engineers (1993)
220. Miyakoshi, K., Okada, H.: Estimation of the site response in the Kushiro city, Hokkaido, Japan, using microtremors with seismometer arrays. In: 10th World Conference on Earthquake Engineering (1996)

221. Mochizuki, Y., Fukushima, S., Kagawa, K.: Shaking table test on reinforced sand. In: Proceedings of Geotextile Symposium, vol. 2, pp. 9–13. Japan Chapter of International Geosynthetics Society, Tokyo (1987)
222. Mohanty, W., Walling, M.: Seismic hazard in mega city Kolkata, India. Nat. Hazards **47**(1), 39–54 (2008)
223. Mohanty, W., Walling, M., Nath, S., Pal, I.: First order seismic microzonation of Delhi, India using geographic information system (GIS). Nat. Hazards **40**(2), 245–260 (2007)
224. Morales, E.M.: Liquefaction-induced damage to buildings in 1990 Luzon earthquake. J. Geotech. Eng. **121**(5), 453–454 (1995)
225. Moss, R.E., Seed, R.B., Kayen, R.E., Stewart, J.P., Der Kiureghian, A., Cetin, K.O.: CPT-based probabilistic and deterministic assessment of in situ seismic soil liquefaction potential. J. Geotech. Geoenviron. Eng. **132**(8), 1032–1051 (2006)
226. Moyle, W. Jr.: Ground-water-level monitoring for earthquake prediction; a progress report based on data collected in southern California, 1976–79. Tech. rep., Water Resources Division, US Geological Survey (1980)
227. Mukhopadhyay, S., Pandey, Y., Dharmaraju, R., Chauhan, P., Singh, P., Dev, A.: Seismic microzonation of Delhi for ground-shaking site effects. Curr. Sci. **82**(7), 877–881 (2002)
228. Naein, S., Moayed, R.Z.: Improving liquefaction potential strength by using micropiles. IAEG the Geological Society of London, pp. 333–338 (2006)
229. Nakamura, Y.: A method for dynamic characteristics estimation of subsurface using microtremor on the ground surface, vol. 30(1). Railway Technical Research Institute, Quarterly Reports (1989)
230. Nakazawa, H., Ishihara, K., Tsukamoto, Y., Kamata, T.: Case studies on evaluation of liquefaction resistance of imperfectly saturated soil deposits. In: Proceedings of International Conference on Cyclic Behaviour of Soils and Liquefaction Phenomena, Bochum, vol. 31, pp. 295–304 (2004)
231. Narayan, J., Sharma, M.: Effects of local geology on damage severity during Bhuj, India earthquake. In: 13th World Conference on Earthquake Engineering, Vancouver (2004)
232. Nath, S., Thingbaijam, K.: Probabilistic seismic hazard assessment of India. Seismol. Res. Lett. **83**(1), 135–149 (2012)
233. Nath, S., Chatterjee, D., Biswas, N., Dravinski, M., Cole, D., Papageorgiou, A., Rodriguez, J., Poran, C.: Correlation study of shear wave velocity in near surface geological formations in anchorage, Alaska. Earthq. Spectra **13**(1), 55–75 (1997)
234. Nath, S., Vyas, M., Pal, I., Singh, A., Mukherjee, S., Sengupta, P.: Spectral attenuation models in the Sikkim Himalaya from the observed and simulated strong motion events in the region. Curr. Sci. **88**(2), 295–303 (2005)
235. Nath, S.K., Thingbaijam, K.K.S., Raj, A.: Earthquake hazard in northeast India-a seismic microzonation approach with typical case studies from Sikkim Himalaya and Guwahati city. J. Earth Syst. Sci. **117**(2), 809–831 (2008)
236. Nath, S., Raj, A., Thingbaijam, K., Kumar, A.: Ground motion synthesis and seismic scenario in Guwahati city a stochastic approach. Seismol. Res. Lett. **80**(2), 233–242 (2009)
237. NDMA: Development of Probabilistic Seismic Hazard Map of India. National Disaster Management Authority, Government of India, New Delhi (2010)
238. NDMA: Geotechnical/Geophysical Investigations for Seismic Microzonation Studies of Urban Centres in India. National Disaster Management Authority, Government of India, New Delhi (2011)
239. Neelima Satyam, D., Rao, K.: Microtremor studies for seismic site characterization of Delhi region. In: The 12th International Conference of International Association for Computer Methods and Advances in Geomechanics (IACMAG), Goa, pp. 2811–2815 (2008)
240. Fäh, D., Rüttener, E., Noack, T., Kruspan, P.: Microzonation of the city of Basel. J. Seismol. **1**(1), 87–102 (1997)
241. Nuttli, O.W.: Average seismic source-parameter relations for mid-plate earthquakes. Bull. Seismol. Soc. Am. **73**(2), 519–535 (1983)

242. Ohba, S., Toriumi, I.: Dynamic response characteristics of Osaka plain. In: Proceedings of the Annual Meeting AIJ (1970)
243. Ohmachi, T., Nakamura, Y., Toshinawa, T.: Ground motion characteristics in the San Francisco bay area detected by microtremor measurements. In: Proceedings of the Second International Conference on Recent Advances in Geotechnical Earthquake Engineering and Soil Dynamics, March, pp. 11–15 (1991)
244. Ohsaki, Y., Iwasaki, R.: On dynamic shear moduli and Poisson's ratios of soil deposits. Soils Found. **13**(4), 61–73 (1973)
245. Ohta, Y., Goto, N.: Empirical shear wave velocity equations in terms of characteristic soil indexes. Earthq. Eng. Struct. Dyn. **6**(2), 167–187 (1978)
246. Okal, E.A.: Use of the mantle magnitude mm for the reassessment of the moment of historical earthquakes. Pure Appl. Geophys. **139**(1), 17–57 (1992)
247. Okal, E.A., Reymond, D.: The mechanism of great Banda Sea earthquake of 1 February 1938: applying the method of preliminary determination of focal mechanism to a historical event. Earth Planet. Sci. Lett. **216**(1–2), 1–15 (2003)
248. Okamoto, T., Kokusho, T., Yoshida, Y., Kusuonoki, K.: Comparison of surface versus subsurface wave source for p-s logging in sand layer. In: Proceedings of 44th Annual Conference of the JSCE, vol. 3, pp. 996–997 (1989)
249. Okamura, M., Teraoka, T.: Shaking table tests to investigate soil desaturation as a liquefaction countermeasure. In: Seismic Performance and Simulation of Pile Foundations in Liquefied and Laterally Spreading Ground, pp. 282–293. American Society of Civil Engineers, Reston (2006)
250. Oldham, T.: A catalogue of Indian earthquakes ed. by R.D. Oldham. Mem. Geol. Surv. India **19**, 163–215 (1883)
251. O'Neill, A.: Some pitfalls associated with dominant higher-mode inversion. In: Proceedings of the 8th International Symposium on Recent Advances in Exploration Geophysics, Kyoto University, pp. 48–55 (2004)
252. Ono, Y., Nakajima, Y., Ito, K., Oishi, H.: Efficient installation of gravel drains. In: International Conferences on Recent Advances in Geotechnical Earthquake Engineering and Soil Dynamics, 18 (1991)
253. Ordaz, M., Aguilar, A., Arboleda, J.: Crisis2007: program for computing seismic hazard. Instituto de Ingenierıa, Universidad Naconal Autónoma de México, UNAM, México (2007)
254. Ordonez, G.: SHAKE-2000, a Computer Program for the 1-D Analysis of Geotechnical Earthquake Engineering Problems, User's Manual (2000)
255. Panza, G., Vaccari, F., Cazzaro, R.: Deterministic seismic hazard assessment. In: Vrancea Earthquakes: Tectonics, Hazard and Risk Mitigation, pp. 269–286. Kluwer Academic Publishers, Dordrecht (1999)
256. Papazachos, B., Karakostas, V., Kiratzi, A., Margaris, B., Papazachos, C., Scordilis, E.: Uncertainties in the estimation of earthquake magnitudes inGreece. J. Seismol. **6**(4), 557–570 (2002)
257. Park, C., Miller, R., Xia, J.: Imaging dispersion curves of surface waves on multi-channel record. In: 1998 SEG Annual Meeting (1998)
258. Park, C., Miller, R., Xia, J.: Multichannel analysis of surface waves MASW. Geophysics **64**(3), 800–808 (1999)
259. Park, C., Heljeson, M., Ivanov, J., Brohammer, M.: Surfseis User's Manual v 2.0. Kansas Geological Survey, Kansas (2007)
260. Parvez, I., Vaccari, F., Panza, G.: A deterministic seismic hazard map of India and adjacent areas. Geophys. J. Int. **155**(2), 489–508 (2003)
261. Parvez, I., Vaccari, F., Panza, G.: Site-specific microzonation study in Delhi metropolitan city by 2-D modelling of SH and P-SV waves. Pure Appl. Geophys. **161**, 1165–1184 (2004)
262. Patton, H.J., Walter, W.R.: Regional moment: magnitude relations for earthquakes and explosions. Geophys. Res. Lett. **20**(4), 277–280 (1993)

263. PCRSMJUA: Project Completion Report of Seismic Microzonation of Jabalpur Urban Area. Department of Science and Technology, Government of India (2005)
264. Pedersen, H., Le Brun, B., Hatzfeld, D., Campillo, M., Bard, P.Y.: Ground-motion amplitude across ridges. Bull. Seismol. Soc. Am. **84**(6), 1786–1800 (1994)
265. Phillips, W.S., Aki, K.: Site amplification of coda waves from local earthquakes in central California. Bull. Seismol. Soc. Am. **76**(3), 627–648 (1986)
266. Pitilakis, K.: Site effects. In: Recent Advances in Earthquake Geotechnical Engineering and Microzonation, pp. 139–197. Springer, Dordrecht (2004)
267. Poulos, S.J.: The Stress-Strain Curves of Soils. Geotechnical Engineers Incorporated, Winchester (1971)
268. Poulos, S.J.: The steady state of deformation. J. Geotech. Geoenviron. Eng. **107**, 553–562 (1981). ASCE 16241 Proceeding
269. Power, M., Chiou, B., Abrahamson, N., Bozorgnia, Y., Shantz, T., Roblee, C.: An overview of the NGA project. Earthq. Spectra **24**(1), 3–21 (2008)
270. Prevost, J.: DYNA1D: A Computer Program for Nonlinear Seismic Site Response Analysis Technical Documentation. National Center for Earthquake Engineering Research, Buffalo (1989)
271. Promputthangkoon, P., Hyde, A.: Compressibility and liquefaction potential of rubber composite soils. In: Scrap Tire Derived Geomaterials-Opportunities and Challenges: Proceedings of the International Workshop IW-TDGM 2007 (Yokosuka, 23–24 March 2007), p. 161. CRC Press, Boca Raton (2007)
272. Pyke, R., Chan, C., Seed, H.: Settlement and Liquefaction of Sands Under Multi-Directional Shaking. Earthquake Engineering Research Center, University of California, Berkeley (1974)
273. Raghu Kanth, S., Iyengar, R.: Seismic hazard estimation for Mumbai city. Curr. Sci. **91**(11), 1486–1494 (2006)
274. Raghu Kanth, S., Iyengar, R.: Estimation of seismic spectral acceleration in peninsular India. J. Earth Syst. Sci. **116**(3), 199–214 (2007)
275. Rao, B.: Historical seismicity and deformation rates in the Indian peninsular shield. J. Seismol. **4**(3), 247–258 (2000)
276. Rao, K., Mohanty, W.: The Bhuj earthquake and lessons for the damages. IGS News **33**(2), 3–10 (2001)
277. Rao, K., Neelima Satyam, D.: Seismic microzonation studies for Delhi region. In: Symposium on Seismic Hazard Analysis and Microzonation, September 23–24, Roorkee (2005)
278. Rao, B.R., Rao, P.S.: Historical seismicity of peninsular India. Bull. Seismol. Soc. Am. **74**(6), 2519–2533 (1984)
279. Raptakis, D., Anastasiadis, A., Pitilakis, K., Lontzetidis, K.: Shear wave velocities and damping of Greek natural soils. In: Proceedings of the 10th European Conference on Earthquake Engineering, vol. 1, pp. 477–482 (1994)
280. Rastogi, B.: Earthquake mechanisms and plate tectonics in the Himalayan region. Tectonophysics **21**(1–2), 47–56 (1974)
281. Reid, H.: The California Earthquake of April 18, 1906, vol. 27. Carnegie Institution, Washington (1910)
282. Reiter, L.: Earthquake Hazard Analysis: Issues and Insights. Columbia University Press, New York (1990)
283. Richter, C.: An instrumental earthquake magnitude scale. Bull. Seismol. Soc. Am. **25**(1), 1–32 (1935)
284. Riepl, J., Bard, P.Y., Hatzfeld, D., Papaioannou, C., Nechtschein, S.: Detailed evaluation of site-response estimation methods across and along the sedimentary valley of Volvi (Euroseistest). Bull. Seismol. Soc. Am. **88**(2), 488–502 (1998)
285. Robertson, P.: Soil classification using the cone penetration test. Can. Geotech. J. **27**(1), 151–158 (1990)

286. Robertson, P., Campanella, R.: Liquefaction potential of sands using the CPT. J. Geotech. Eng. **111**(3), 384–403 (1985)
287. Robertson, P., Wride, C.: Cyclic liquefaction and its evaluation based on the SPT and CPT. In: Proceedings of the NCEER Workshop on Evaluation of Liquefaction Resistance of Soils. Technical Report NCEER-97-0022, National Center for Earthquake Engineering Research, Buffalo, pp. 41–87 (1997)
288. Robertson, P., Wride, C.: Evaluating cyclic liquefaction potential using the cone penetration test. Can. Geotech. J. **35**(3), 442–459 (1998)
289. Robertson, P., Campanella, R., Wightman, A.: SPT-CPT correlations. J. Geotech. Eng. **109**(11), 1449–1459 (1983)
290. Rodriguez-Marek, A., Bray, J.D., Abrahamson, N.A.: An empirical geotechnical seismic site response procedure. Earthq. Spectra **17**(1), 65–87 (2001)
291. Rogers, J.: Subsurface exploration using the standard penetration test and the cone penetrometer test. Environ. Eng. Geosci. **12**(2), 161–179 (2006)
292. Rollins, K., Evans, M., Diehl, N., William D. III: Shear modulus and damping relationships for gravels. J. Geotech. Geoenviron. Eng. **124**(5), 396–405 (1998)
293. Rugg, D., Yoon, J., Hwang, H., El Mohtar, C.: Undrained shearing properties of sand permeated with a bentonite suspension for static liquefaction mitigation. In: Geo-Frontiers 2011: Advances in Geotechnical Engineering, pp. 677–686. American Society of Civil Engineers, Reston (2011)
294. Rydelek, P.A., Sacks, I.S.: Testing the completeness of earthquake catalogues and the hypothesis of self-similarity. Nature **337**(6204), 251–253 (1989)
295. Saaty, T.: The Analytic Hierarchy Process. McGraw-Hill, New York (1980)
296. Sadigh, K., Chang, C., Egan, J., Makdisi, F., Youngs, R.: Attenuation relationships for shallow crustal earthquakes based on California strong motion data. Seismol. Res. Lett. **68**(1), 180–189 (1997)
297. Sánchez-Sesma, F.J.: Diffraction of elastic waves by three-dimensional surface irregularities. Bull. Seismol. Soc. Am. **73**(6A), 1621–1636 (1983)
298. Sánchez-Sesma, F.J.: Diffraction of elastic SH waves by wedges. Bull. Seismol. Soc. Am. **75**(5), 1435–1446 (1985)
299. Sánchez-Sesma, F., Chávez-Pérez, S., Suarez, M., Bravo, M., Pérez-Rocha, L.: The méxico earthquake of 19 September 1985 on the seismic response of the valley of méxico. Earthq. Spectra **4**(3), 569–589 (1988)
300. Sasaki, Y., Taniguchi, E.: Shaking table tests on gravel drains to prevent liquefaction of sand deposits. Soils Found. **22**(3), 1–14 (1982)
301. Schnabel, P.: Effects of local geology and distance from source on earthquake ground motions. University of California, Berkeley (1973)
302. Schnabel, P., Lysmer, J., Seed, H.: SHAKE: a computer program for earthquake response analysis of horizontally layered sites. Earthquake Engineering Research Center. Report No EERC, pp. 72–12 (1972)
303. Scordilis, E.: Empirical global relations converting M_s and m_b to moment magnitude. J. Seismol. **10**(2), 225–236 (2006)
304. Seeber, L., Armbruster, J., Jacob, K.: Probabilistic assessment of seismic hazard for Maharashtra. Government of Maharashtra, Unpublished Report (1999)
305. Seed, H.B., Booker, J.R.: Stabilization of potentially liquefiable sand deposits using gravel drains. J. Geotech. Geoenviron. Eng. **103**(ASCE 13050) (1977)
306. Seed, H., De Alba, P.: Use of SPT and CPT tests for evaluating the liquefaction resistance of sands. Use of in Situ Tests in Geotechnical Engineering, pp. 281–302. ASCE (1986)
307. Seed, H., Idriss, I.: Influence of soil conditions on ground motions during earthquakes. University of California, Institute of Transportation and Traffic Engineering, Soil Mechanics Laboratory (1969)
308. Seed, H., Idriss, I.: Soil moduli and damping factors for dynamic response analyses. Report No. EERC 70-10, University of California, Berkeley (1970)

309. Seed, H., Idriss, I.: Simplified procedure for evaluating soil liquefaction potential. J. Soil Mech. Found. Div. **97**(9), 1249–1273 (1971)
310. Seed, H., Idriss, I.: Evaluation of liquefaction potential sand deposits based on observation of performance in previous earthquakes. ASCE National Convention (MO), pp. 481–544 (1981)
311. Seed, B., Lee, K.L.: Liquefaction of saturated sands during cyclic loading. J. Soil Mech. Found. Div. **92**, 105–134 (1966). ASCE# 4972 Proceeding
312. Seed, H., Sun, J.: Implication of site effects in the Mexico city earthquake of 19 September 1985 for earthquake-resistance-design criteria in the San Francisco bay area of California. Report No. UCB/EERC-89/03, University of California, Berkeley (1989)
313. Seed, H., Idriss, I., Arango, I.: Evaluation of liquefaction potential using field performance data. J. Geotech. Eng. **109**(3), 458–482 (1983)
314. Seed, H., Tokimatsu, K., Harder, L., Chung, R.: The influence of SPT procedures in soil liquefaction resistance evaluations: Berkeley, University of California. Earthquake Engineering Research Center Report UBC/EERC-84/15 (1984)
315. Seed, H., Tokimatsu, K., Harder, L., Chung, R.: Influence of SPT procedures in soil liquefaction resistance evaluations. J. Geotech. Eng. **111**(12), 1425–1445 (1985)
316. SEISAT: Seismotectonic Atlas of India. Geological Survey of India, New Delhi (2000)
317. Semblat, J.F., Duval, A.M., Dangla, P.: Numerical analysis of seismic wave amplification in nice (France) and comparisons with experiments. Soil Dyn. Earthq. Eng. **19**(5), 347–362 (2000)
318. Sharma, M.L.: Attenuation relationship for estimation of peak ground horizontal acceleration using data from strong-motion arrays in India. Bull. Seismol. Soc. Am. **88**(4), 1063–1069 (1998)
319. Sharma, M., Malik, S.: Probabilistic seismic hazard analysis and estimation of spectral strong ground motion on bed rock in North East India. In: Fourth International Conference on Earthquake Engineering, Taipei, Paper, 15 (2006)
320. Sharma, M., Wason, H., Dimri, R.: Seismic zonation of the Delhi region for bedrock ground motion. Pure Appl. Geophys. **160**(12), 2381–2398 (2003)
321. Sharma, B., Gupta, A.K., Devi, D.K., Kumar, D., Teotia, S., Rastogi, B.: Attenuation of high-frequency seismic waves in Kachchh region, Gujarat, India. Bull. Seismol. Soc. Am. **98**(5), 2325–2340 (2008)
322. Sharma, M., Douglas, J., Bungum, H., Kotadia, J.: Ground-motion prediction equations based on data from the Himalayan and Zagros regions. J. Earthq. Eng. **13**(8), 1191–1210 (2009)
323. Shibata, T.: Analysis of liquefaction of saturated sand during cyclic loading. Disaster Prev. Res. Inst. Bull. **13**, 563–570 (1970)
324. Shima, E.: Seismic microzoning map of Tokyo. In: Proceedings of the Second International Conference on Microzonation, vol. 1, pp. 433–443 (1978)
325. Shtivelman, V.: Surface wave sections as a tool for imaging subsurface inhomogeneities. Eur. J. Environ. Eng. Geophys. **7**, 121–138 (2002)
326. Singh, R.: Probabilistic seismic hazard and risk analysis: a case study for Ahmedabad city. Ph.D. thesis, Indian Institute of Technology, Kanpur (2009)
327. Sitharam, T., Anbazhagan, P.: Seismic hazard analysis for the Bangalore region. Nat. Hazards **40**(2), 261–278 (2007)
328. Sitharam, T., Anbazhagan, P.: Seismic microzonation: principles, practices and experiments. EJGE Special Volume Bouquet 8, 61 (2008)
329. Sitharam, T., Anbazahgan, P.: Report on Seismic Microzonation of Bangalore Urban Centre (Vol. 1). Seismology Division, Ministry of Earth Sciences Government of India, New Delhi (2009)
330. Sitharam, T., Dinesh, S.: Numerical simulation of liquefaction behaviour of granular materials using discrete element method. J. Earth Syst. Sci. **112**(3), 479–484 (2003)
331. Sitharam, T., Kolathayar, S.: Seismic hazard analysis of India using areal sources. J. Asian Earth Sci. **62**, 647–653 (2013)
332. Sitharam, T., Sil, A.: Comprehensive seismic hazard assessment of Tripura and Mizoram states. J. Earth Syst. Sci. **123**(4), 837–857 (2014)

333. Sitharam, T., Anbazhagan, P., Ganesha Raj, K.: Use of remote sensing and seismotectonic parameters for seismic hazard analysis of Bangalore. Nat. Hazards Earth Syst. Sci. **6**(6), 927–939 (2006)
334. Sitharam, T., James, N., Vipin, K., Raj, K.: A study on seismicity and seismic hazard for Karnataka state. J. Earth Syst. Sci. **121**(2), 475–490 (2012)
335. Sitharam, T., Kolathayar, S., James, N.: Probabilistic assessment of surface level seismic hazard in India using topographic gradient as a proxy for site condition. Geosci. Front. **6**(6), 847–859 (2015)
336. Skempton, A.: Standard penetration test procedures and the effects in sands of overburden pressure, relative density, particle size, ageing and over consolidation. Geotechnique **36**(3), 425–447 (1986)
337. Sladen, J.A., D'hollander, R.D., Krahn, J.: The liquefaction of sands, a collapse surface approach. Can. Geotech. J. **22**(4), 564–578 (1985)
338. Slob, S., Hack, R., Scarpas, T., van Bemmelen, B., Duque, A.: A methodology for seismic microzonation using GIS and Shakea case study from Armenia, Colombia. In: Engineering Geology for Developing Countries – Proceedings of 9th Congress of the International Association for Engineering Geology and the Environment, Durban, pp. 16–20 (2002)
339. Sonmez, H.: Modification of the liquefaction potential index and liquefaction susceptibility mapping for a liquefaction-prone area (Inegol, Turkey). Environ. Geol. **44**(7), 862–871 (2003)
340. Sonu, C.J., Ito, K., Oishi, H.: Harry seed, liquefaction and the gravel drain. Civil Eng. **63**(12), 58 (1993)
341. Srivastava, H., Ramachandran, K.: New catalogue of earthquakes for peninsular India during 1839–1900. Mausam **36**(3), 351–358 (1985)
342. Stepp, J.: Analysis of completeness of the earthquake sample in the Puget Sound area and its effect on statistical estimates of earthquake hazard. In: Proceedings of the International Conference on Microzonation for Safer Construction Research and Application, vol. 64(4), 1189–1207 (1972)
343. Stepp, J., Wong, I., Whitney, J., Quittmeyer, R., Abrahamson, N., Toro, G., Youngs, R., Coppersmith, K., Savy, J., Sullivan, T.: Probabilistic seismic hazard analyses for ground motions and fault displacement at Yucca mountain, Nevada. Earthq. Spectra **17**(1), 113–151 (2001)
344. Stewart, J., Choi, Y., Liu, A., Baturay, M.: Amplification factors for spectral acceleration in active regions. Report No. PEER-2001/10, Pacific Earthquake Engineering Research Center, University of California, Berkeley, vol. 2001 (2001). http://www.cee.ucla.edu/Faculty/jstewart/publications.htm
345. Stewart, J., Liu, A., Choi, Y.: Amplification factors for spectral acceleration in tectonically active regions. Bull. Seismol. Soc. Am. **93**(1), 332–352 (2003)
346. Stokoe, K., Roesset, J., Bierschwale, J., Aouad, M.: Liquefaction potential of sands from shear wave velocity. In: Proceedings, 9nd World Conference on Earthquake, vol. 13, pp. 213–218 (1988)
347. Stokoe, K.H., Rix, G.J., Nazarian, S.: In situ seismic testing with surface wave: processing. In: XII International Conference on Soil Mechanics and Foundation Engineering, vol. 33, pp. 1–334 (1989)
348. Stone, W., Yokel, F., Celebi, M., Hanks, T., Leyendecker, E.: Engineering aspects of the 19 September 1985 Mexico earthquake. US Department of Commerce, National Bureau of Standards, Washington (1987)
349. Sykora, D.W.: Correlations of in situ measurements in sands of shear wave velocity, soil characteristics, and site conditions. Ph.D. thesis, University of Texas at Austin (1983)
350. Sykora, D.W., Koester, J.P.: Correlations between dynamic shear resistance and standard penetration resistance in soils. In: Earthquake Engineering and Soil Dynamics II: Recent Advances in Ground-Motion Evaluation, pp. 389–404. American Society of Civil Engineers, New York (1988)
351. Tandon, A., Srivastava, H.: The stress drop and average dislocation of some earthquakes in the Indian sub-continent. Pure Appl. Geophys. **112**(6), 1051–1057 (1974)

352. Theodulidis, N., Bard, P.Y.: Horizontal to vertical spectral ratio and geological conditions: an analysis of strong motion data from Greece and Taiwan (smart-1). Soil Dyn. Earthq. Eng. **14**(3), 177–197 (1995)
353. Theodulidis, N., Bard, P.Y., Archuleta, R., Bouchon, M.: Horizontal-to-vertical spectral ratio and geological conditions: the case of Garner Valley Downhole Array in southern California. Bull. Seismol. Soc. Am. **86**(2), 306–319 (1996)
354. Thingbaijam, K.K.S., Nath, S.K., Yadav, A., Raj, A., Walling, M.Y., Mohanty, W.K.: Recent seismicity in northeast India and its adjoining region. J. Seismol. **12**(1), 107–123 (2008)
355. Tokimatsu, K., Kuwayama, S., Tamura, S.: Liquefaction potential evaluation based on Rayleigh wave investigation and its comparison with field behavior. In: Proceedings, pp. 357–364 (1991)
356. Tokimatsu, K., Nakajo, Y., Tamura, S.: Horizontal to vertical amplitude ratio of short period microtremors and its relation to site characteristics. J. Struct. Construction Eng. **457**, 11–18 (1994)
357. Topal, T., Doyuran, V., Karahanoğlu, N., Toprak, V., Süzen, M., Yeşilnacar, E.: Microzonation for earthquake hazards: Yenişehir settlement, Bursa, Turkey. Eng. Geol. **70**(1), 93–108 (2003)
358. Toprak, S., Holzer, T., Bennett, M., Tinsley, J.: CPT-and SPT-based probabilistic assessment of liquefaction potential. In: Proceedings of the 7th US–Japan Workshop on Earthquake Resistant Design of Lifeline Facilities and Countermeasures Against Liquefaction (1999)
359. Toro, G., Abrahamson, N., Schneider, J.: Model of strong ground motions from earthquakes in central and eastern North America: best estimates and uncertainties. Seismol. Res. Lett. **68**(1), 41–57 (1997)
360. Towhata, I.: Geotechnical Earthquake Engineering. Springer, Dordrecht (2008)
361. Trifunac, M.D., Hudson, D.E.: Analysis of the pacoima dam accelerogram – San Fernando, California, earthquake of 1971. Bull. Seismol. Soc. Am. **61**(5), 1393–1411 (1971)
362. Trifunac, M., Todorovska, M.: Nonlinear soil response as a natural passive isolation mechanism – the 1994 Northridge, California, earthquake. Soil Dyn. Earthq. Eng. **17**(1), 41–51 (1998)
363. Tsuchida, H.: Evaluation of liquefaction potential of sandy deposits and measures against liquefaction induced damage. In: Proceedings of the Annual Seminar of the Port and Harbour Research Institute, pp.(3-1)–(3-33) (1970; in Japanese)
364. Tsukamoto, Y., Ishihara, K., Nakazawa, H., Kamada, K., Huang, Y.: Resistance of partly saturated sand to liquefaction with reference to longitudinal and shear wave velocities. Soils Found. **42**(6), 93–104 (2002)
365. Uchimura, T., Chi, N., Nirmalan, S., Sato, T., Meidani, M., Towhata, I.: Shaking table tests on effect of tire chips and sand mixture in increasing liquefaction resistance and mitigating uplift of pipe. In: Proceedings, International Workshop on Scrap Tire Derived Geomaterials – Opportunities and Challenges, Yokosuka, pp. 179–186 (2007)
366. Udwadia, F., Trifunac, M.: Comparison of earthquake and microtremor ground motions in El centro, California. Bull. Seismol. Soc. Am. **63**(4), 1227–1253 (1973)
367. Uma Maheswari, R., Boominathan, A., Dodagoudar, G.: Seismic site classification and site period mapping of Chennai city using geophysical and geotechnical data. J. Appl. Geochem. **72**(3), 152–168 (2010)
368. USGS: Virginia Well Records Sumatra-Andaman Islands Earthquake. USGS News Release (2005)
369. Utsu, T.: A method for determining the value of b in formula log n = a-bm showing the magnitude-frequency relation for earthquakes. Geophys. Bull. Hokkaido Univ. **13**, 99–103 (1965)
370. Vaid, Y.P., Chern, J.C.: Effect of static shear on resistance to liquefaction. Soils Found. **23**(1), 47–60 (1983)
371. Vaid, Y., Chern, J.: Effect of static shear on resistance to liquefaction. Soils Found. **25**(3), 154–156 (1985)

372. Van Paassen, L., Harkes, M., Van Zwieten, G., Van der Zon, W., Van der Star, W., Van Loosdrecht, M.: Scale up of Biogrout: a biological ground reinforcement method. In: Proceedings of the 17th International Conference on Soil Mechanics and Geotechnical Engineering, pp. 2328–2333. IOS Press, Lansdale (2009)
373. Vilanova, S., Fonseca, J.: Probabilistic seismic-hazard assessment for Portugal. Bull. Seismol. Soc. Am. **97**(5), 1702–1717 (2007)
374. Vipin, K.S., Sitharam, T.G., Anbazhagan, P.: Probabilistic evaluation of seismic soil liquefaction potential based on SPT data. Nat. Hazards **53**(3), 547–560 (2010)
375. Vipin, K.S., Sitharam, T.G.: A performance-based framework for assessing liquefaction potential based on CPT data. Georisk: Assess. Manag. Risk Eng. Syst. Geohazards **6**(3), 177–187 (2012)
376. Vipin, K., Anbazhagan, P., Sitharam, T.: Estimation of peak ground acceleration and spectral acceleration for south India with local site effects: probabilistic approach. Nat. Hazards Earth Syst. Sci. **9**, 865–878 (2009)
377. Vucetic, M., Dobry, R.: Effect of soil plasticity on cyclic response. J. Geotech. Eng. **117**(1), 89–107 (1991)
378. Wahlström, R., Grünthal, G.: Probabilistic seismic hazard assessment (horizontal PGA) for Sweden, Finland and Denmark using different logic tree approaches. Soil Dyn. Earthq. Eng. **20**(1), 45–58 (2000)
379. Wald, D., Allen, T.: Topographic slope as a proxy for seismic site conditions and amplification. Bull. Seismol. Soc. Am. **97**(5), 1379–1395 (2007)
380. Wang, W.: Some Findings in Soil Liquefaction Earthquake Engineering. Department Water Conservancy and Hydroelectric Power Scientific Research Institute, Beijing (1979)
381. Wiemer, S., Wyss, M.: Minimum magnitude of completeness in earthquake catalogs: Examples from Alaska, the western united states, and Japan. Bull. Seismol. Soc. Am. **90**(4), 859–869 (2000)
382. Wiemer, S., Wyss, M.: Mapping spatial variability of the frequency-magnitude distribution of earthquakes. Adv. Geophys. **45**, 259–302 (2002)
383. Wills, C.J., Silva, W.: Shear-wave velocity characteristics of geologic units in California. Earthq. Spectra **14**(3), 533–556 (1998)
384. Wills, C., Petersen, M., Bryant, W., Reichle, M., Saucedo, G., Tan, S., Taylor, G., Treiman, J.: A site-conditions map for California based on geology and shear-wave velocity. Bull. Seismol. Soc. Am. **90**(6B), S187–S208 (2000)
385. Woessner, J., Wiemer, S.: Assessing the quality of earthquake catalogues: estimating the magnitude of completeness and its uncertainty. Bull. Seismol. Soc. Am. **95**(2), 684 (2005). http://dx.doi.org/10.1785/0120040007
386. Woo, G.: Kernel estimation methods for seismic hazard area source modeling. Bull. Seismol. Soc. Am. **86**(2), 353–362 (1996)
387. Wood, H.: Distribution of apparent intensity in San Francisco. Report of the State Earthquake Investigation Commission 1(1906), pp. 220–245 (1908)
388. Xia, J., Miller, R., Park, C.: Estimation of near-surface shear-wave velocity by inversion of Rayleigh waves. Geophysics **64**(3), 691–700 (1999)
389. Xia, J., Miller, R.D., Park, C.B., Hunter, J.A., Harris, J.B., Ivanov, J.: Comparing shear-wave velocity profiles inverted from multichannel surface wave with borehole measurements. Soil Dyn. Earthq. Eng. **22**(3), 181–190 (2002)
390. Xu, Z.Y.: Analysis of liquefiable sand deposits using gravel drains. In: International Conferences on Recent Advances in Geotechnical Earthquake Engineering and Soil Dynamics, 6 (1991)
391. Yanagisawa, E.: Damage to structures due to liquefaction in the Japan sea earthquake of 1983. Disasters **7**(4), 259–265 (1983)
392. Yang, J.: Non-uniqueness of flow liquefaction line for loose sand. Geotechnique Lond. **52**(10), 757–760 (2002)
393. Yang, J., Savidis, S., Roemer, M.: Evaluating liquefaction strength of partially saturated sand. J. Geotech. Geoenviron. Eng. **130**(9), 975–979 (2004)

394. Yasuhara, K., Komine, H., Murakami, S., Miyota, S., Hazarika, H.: Mitigation of liquefaction using tire chips as a gravel drain. In: Proceedings of the 6th International Congress on Environmental Geotechnology. New Delhi, pp. 1176–1181 (2010)
395. Yeats, R., Sieh, K., Allen, C.: The Geology of Earthquakes, vol. 568. Oxford University Press, New York (1997)
396. Yegian, M., Eseller-Bayat, E., Alshawabkeh, A., Ali, S.: Induced-partial saturation for liquefaction mitigation: experimental investigation. J. Geotech. Geoenviron. Eng. **133**(4), 372–380 (2007)
397. Yoshida, N., Suetomi, I.: DYNEQ: a computer program for dynamic analysis of level ground based on equivalent linear method. Reports of Engineering Research Institute, Sato Kogyo, pp. 61–70 (1996)
398. Yoshimi, Y., Kuwabara, F.: Effect of subsurface liquefaction on the strength of surface soil. Soils Found. **13**(2), 67–81 (1973)
399. Youd, T., Noble, S., Idriss, I.: Liquefaction criteria based on statistical and probabilistic analyses. Tech. Rep. NCEER 97, 201–15 (1997)
400. Youd, T., Idriss, I., Andrus, R., Arango, I., Castro, G., Christian, J., Dobry, R., W.D., L.F., Harder, L., Hynes, M., Ishihara, K., Koester, J., Laio, S., Marcuson, W., Martin, G., Mitchell, J., Moriwaki, Y., Power, M., Robertson, P., Seed, R., Stokoe, K.: Liquefaction resistance of soils. Summary report from the 1996 NCEER and 1998 NCEER/NSF Workshops on Evaluation of Liquefaction Resistance of Soils. J. Geotech. Geoenviron. Eng. **127**(10), 817–833 (2001)
401. Yu, Y.x., Wang, S.y.: Attenuation relations for horizontal peak ground acceleration and response spectrum in northeastern Tibetan plateau region. Acta Seismol. Sin. **17**(6), 651–661 (2004)
402. Zare, M., Bard, P.Y., Ghafory-Ashtiany, M.: Site characterizations for the Iranian strong motion network. Soil Dyn. Earthq. Eng. **18**(2), 101–123 (1999)
403. Zhao, F., Zhang, Y.: Artificial ground motion compatible with specified peak velocity and target spectrum. Acta Seismol. Sin. **19**(4), 461–471 (2006)

Index

T. G. Sitharam et al., *Comprehensive Seismic Zonation Schemes for Regions at Different Scales*, https://doi.org/10.1007/978-3-319-89659-5

MIX
Papier aus verantwortungsvollen Quellen
Paper from responsible sources
FSC® C105338

If you have any concerns about our products,
you can contact us on
ProductSafety@springernature.com

In case Publisher is established outside the EU,
the EU authorized representative is:
Springer Nature Customer Service Center GmbH
Europaplatz 3, 69115 Heidelberg, Germany

Printed by Libri Plureos GmbH
in Hamburg, Germany